A S AGENCY

The Driving Standards Agency (I
the Department of the Environm

You'll see its logo at test centres.

DRIVING STANDARDS AGENCY
SAFE DRIVING FOR LIFE

The aim of DSA is to promote road safety through the advancement of driving standards.

DSA

- conducts practical driving tests for motorcylists and drivers of cars, lorries, buses and other vehicles
- plans, maintains and supervises the theory test for motorcyclists and drivers of cars, lorries and buses
- controls the register of Approved Driving Instructors (ADIs)
- supervises Compulsory Basic Training (CBT) for motorcyclists
- controls the voluntary register of Large Goods Vehicle (LGV) instructors
- aims to provide a high-quality service to its customers.

CONTENTS

CONTENTS

helping your business grow more profitably

You may be planning to drive a tractor or other specialist vehicle occasionally, or you may do so regularly as part of your job.

This book is for the driver who needs to take a driving test using a specialist or unusual vehicle. These include three wheeled cars and trikes (Category B1), agricultural tractors (Category F), road rollers (Category G), track laying vehicles (Category H) and mowing machines or pedestrian-controlled vehicles (Category K).

The purpose of the test is to demonstrate basic road safety procedures. It's vitally important to develop the correct attitude towards driving, showing responsibility and consideration to other road users. Only those who can do this will earn the right to drive without L plates (or D plates in Wales).

By passing your test you'll prove that you can drive safely, but it's important that you understand the principles of what you've learned and put them into practice. The instruction you receive before the test is the foundation for gaining further skills and experience. The test is just one stage in your driving career. You shouldn't assume that if you pass your test you're a good driver with nothing more to learn.

During your practical test your examiner will want to see you driving to the standards explained in this book. Those standards are given here in an easy-to-read style with illustrations and guidance for these specialist vehicles. However, driving is never predictable. Road conditions or circumstances coupled with the unusual characteristics of these vehicles will demand that you are able to assess any situation and apply the guidance given in this book.

Make sure that your aim is *Safe Driving for Life.*

Robin Cummins

The Chief Driving Examiner
Driving Standards Agency

This book will help you to

learn to drive competently

prepare and help you to pass your practical driving test.

Part One looks at your driving licence, the types of vehicles you may want to learn to drive, how to prepare yourself and apply for your driving test and what you'll need to bring with you on your test.

Part Two looks at the driving test itself, what your examiner will be looking for and the skills you'll need to show. It also gives advice on the techniques needed to drive specific types of vehicles during your test.

Part Three looks at what you'll need to do after your test. It also explains DSA's complaints and compensation codes.

Part Four looks at using the vehicle on the road.

Part Five lists useful addresses.

The important factors

This book is only one of the important factors in your training.

The others are

- a good instructor
- plenty of practice
- your attitude.

You must manage your own learning.

Aim to be a confident and safe driver for life, not just to pass your test.

Driving is a life skill.

Your test is just the beginning.

Books for study

It's strongly recommended that you study a copy of *The Highway Code* (The Stationery Office). You can buy a copy from most booksellers.

The DSA Driving Skills series of books will provide you with sound knowledge of driving skills and safe practices.

The Official Theory Test for Car Drivers and Motorcyclists, *The Official Driving Manual* and *Know Your Traffic Signs* are published by The Stationery Office.

A CD-ROM, *The Official Theory Test and Beyond* (The Stationery Office) provides a fun and modern way of learning.

PART ONE GETTING STARTED

This part looks at your driving licence, the types of vehicles you may want to learn to drive, how to prepare yourself and apply for your driving test and what you'll need to bring with you on your test.

The topics covered

- Your driving licence.
- Vehicle types.
- Vehicle Excise Duty.
- Insurance.
- Learning to drive.
- The theory test.
- About the driving test.
- Applying for your driving test.
- Attending for your test.
- Your test vehicle.

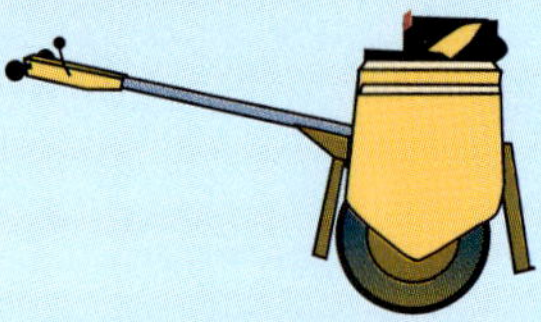

Your driving licence

For the category of vehicle you intend to drive you MUST have either

- a signed, valid GB or Northern Ireland provisional driving licence

or

- a signed valid full driving licence.

The application forms for a driving licence can be obtained from a post office or by contacting the Driver and Vehicle Licencing Agency (DVLA) direct. See Part Five for their address.

When you receive any licence from DVLA you should

- check all the details to make sure they are correct
- return the licence immediately if any of the details are wrong
- when all the details are correct sign the licence in ink, unless it is a photocard licence which will have your signature scanned onto the licence
- keep it in a safe place.

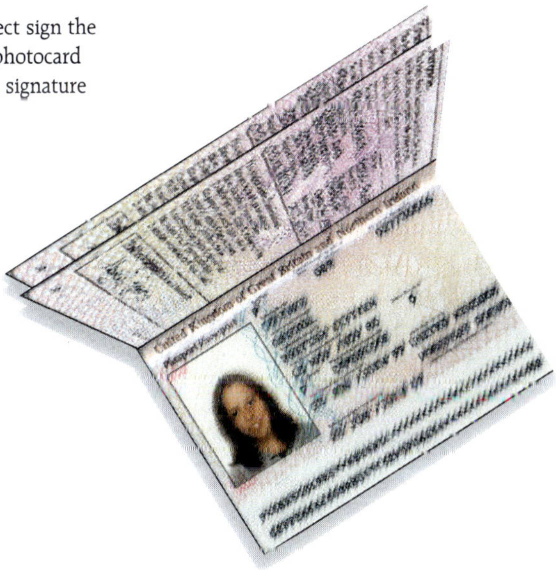

Vehicle types

Category B1 – 3- or 4-wheeled light vehicles

Motor tricycles or quadricycles in category B1 have

- 3 or 4 wheels
- a design speed greater than 50km per hour
- an unladen weight of no more than 550kg.

These vehicles may be designed as

- a small car with car type controls
- a motorcycle with motorcycle type controls.

Age limit

You must be at least 17 years old to use a vehicle in this category on the road.

Speed limit

The speed limits (for cars) listed in *The Highway Code* are relevant to this category of vehicle.

Category F – agricultural tractors

Agricultural or forestry tractors

- have two or more axles
- are constructed for use as a tractor for work off the road in connection with agriculture or forestry.

There are a large variety of agricultural vehicles which come into this category. If you have any doubt you should check with DVLA, whose address is in Part Five.

Age limit

To drive tractors on public roads you must be at least 16 years old. At 16 you're restricted to

- a tractor which is no more than 2.45 metres wide

and only driving on the road

- when travelling to and from a driving test appointment until you've passed your category F driving test.

At 17 years old you can

- drive any tractor
- drive unaccompanied on the road displaying L plates (D plates in Wales).

Speed limit

Agricultural tractors have a speed restriction of

- 20 mph unless the vehicle or any attachment are over 3.5 metres in width, when the limit is reduced to 12 mph
- 40 mph if the vehicle meets constuction and use regulations, including the fitment of
  - safety glass
  - horns
  - additional lights
  - speedometer
  - wings
  - additional noise suppression
  - brakes up to goods vehicle braking standards.

Category G – road rollers

Before you apply for a provisional licence to drive a vehicle in category G you must have passed a test for a vehicle in Category B.

There are two types of road roller you could drive. At 17 years old you can drive a road roller

- with metal rollers
- weighing less than 11.69 tonnes unladen
- which isn't steam propelled.

At 21 years old you can drive other road rollers. These include road rollers

- which have pneumatic, resilient or elastic tyres
- that weigh more than 11.69 tonnes unladen
- that are steam propelled.

Speed limits

Road rollers with resilient tyres are limited to a maximum speed of 20 mph. Road rollers with metal rollers are limited to 5 mph.

Category H – track-laying vehicles steered by their tracks

Before you can apply for a provisional licence to drive a vehicle in category H you must have passed a test for a vehicle in Category B.

This category includes military vehicles, but does not include vehicles designed, constructed and used for excavation work, which are normally allowed to be driven to and from the place of operation by category B licence holders.

Age limit

You must be at least 21 years old to drive these vehicles.

Speed limits

Vehicles with sprung suspension and 'resilient' material between the rims of its weight-carrying rollers and the road surface (this could be rubber or other resilient material fitted as blocks on the tracks or resilient tracks) are limited to 20 mph. Vehicles not so fitted are limited to 5 mph.

Category K – mowing machines or pedestrian-controlled vehicles.

Mowing machines are specialist ride-on grass cutting vehicles with permanent grass cutting equipment.

A pedestrian-controlled vehicle is a powered vehicle where the operator walks with the vehicle, and does not ride on it. (Pedestrian-controlled mowing machines aren't treated as a motor vehicle and so don't need a driving licence).

Age limit

You must be at least 16 years old to drive these vehicles.

Vehicle Excise Duty

This is often called road tax.

The owners or the operators of **all** mechanically-propelled vehicles must display the required Vehicle Excise Licence (tax disc) on the vehicle (Vehicle Excise and Registration Act 1994). A tax disc must not be transferred from one vehicle to another.

Applying

If you're the registered owner, you'll normally receive the form automatically through the post.

If not, you can get a Vehicle Excise Licence Application at any post office.

Most licence-issuing post offices can deal with your application, accept your payment and give you a Vehicle Excise Licence (tax disc).

You can also apply at your nearest Vehicle Registration Office.

DO NOT send the completed application to DVLA.

Duty payable

The rate of vehicle excise duty payable is dependant on the vehicle's

- category
- construction and use
- age.

Contact your local licence-issuing post office or nearest Vehicle Registration Office for the current rates of Vehicle Excise Licence duty. You can pay it either yearly or half-yearly.

Exemptions

All vehicles, except buses and goods vehicles used commercially, are exempt from vehicle excise duty if they were constructed before 1 January 1973.

Insurance

Driving without insurance is

- illegal
- irresponsible.

Should you cause injury to anyone or damage to property, it could be expensive and result in criminal prosecution.

Before you take the vehicle on the public road, make sure you have proper insurance cover.

Third party insurance

This is the legal minimum cover. The 'third party' is any person you might injure or whose property you might damage. You're not covered for damage to your vehicle or injury to yourself.

Third party, fire and theft insurance

The same as third party, except that it also covers against your vehicle being stolen or damaged by fire.

Comprehensive insurance

This is the best, but the most expensive cover. Apart from covering other persons and property from injury and damage, it also covers

- damage to your vehicle
- personal injury to yourself.

What's insured

This varies from company to company. Read the small print and ask the insurer or broker if you're in any doubt.

Insurance companies often don't pay the first £50 or £100 of any claim – this is called the 'excess'.

The certificate of insurance

A short and simple document which certifies

- who's insured
- the vehicle(s) covered
- the kind of cover
- the period of cover
- the main conditions.

Showing your certificate

You'll have to produce the certificate

- if the police ask you
- when renewing the Vehicle Excise Licence
- if you're involved in an accident.

The policy document

This contains the full details of the insurance contract.

It *is* the 'small print', and usually it's written in legal language – although some companies have simplified their policies.

If you don't understand anything, ask your broker or the insurance company to explain it.

Accidents

If you're involved in an accident where

- any person is injured
- any vehicle or property, apart from your own, is damaged
- certain animals not in your vehicle, including large farm animals and dogs, are injured

you must stop.

Give to anyone having reasonable cause to ask for them

- your name and address
- the vehicle owner's name and address if different
- the registration number of the vehicle
- the vehicle make and model
- the name and address of your insurance company
- details of your Insurance Certificate.

If you're not able to do so at the time, you MUST report the accident to the police, in person, as soon as practicable and in any event within 24 hours. If there has been an injury, you must also give insurance details to the police.

If you can't produce the insurance documents when you report the accident, you have up to 7 days to produce them at a police station of your choice.

Learning to drive

An Approved Driving Instructor (ADI) (Car) is approved by DSA to teach learner drivers for payment. Use an ADI if you're learning to drive a vehicle in category B1 with

- controls similar to a car
- two front seats.

It's unlikely that anyone except an ADI would have the experience, knowledge and training to teach you properly.

For other vehicles you should find someone who

- is an experienced driver of the type of vehicle you wish to drive
- can explain clearly and simply the skills you must learn
- is patient and tactful
- builds your confidence.

If your employer needs you to take a test in a specialist vehicle, suitable instruction may be arranged for you.

Practice

When you practise on public roads your vehicle must display L plates to the front and rear (in Wales, you can use D plates if you prefer).

Your practice vehicle must be properly insured for you to drive. It can be useful if your instructor is also insured, so that he or she give you a practical demonstration of the skills you need to learn.

Always make sure your vehicle is roadworthy before using it on the road.

To practise driving a tractor on the road you must be at least 17 years old.

If you are 16 you can drive a tractor on the road.

- when driving to or returning from your driving test appointment
- when you have passed your driving tests.

Where and how to practise

When you're first learning to control your vehicle it's important that you start somewhere quiet with lots of space. Try to find somewhere with

- no pedestrians
- no obstructions
- no moving vehicles
- a good flat surface

and make sure you have the landowner's permission.

Few, if any, specialist vehicles are fitted with dual controls, or even have space for two people on board. Your instructor

- may have to walk alongside you calling out advice
- needs to keep out of your way but be close enough to see what you're doing.

Learn and practise each basic skill before moving on to more complicated techniques. Time spent learning the basics properly will never be wasted.

When you're on the road follow the advice in *The Highway Code.* Always drive on the road unless you're working with your vehicle on a pavement or a grass verge.

Category K vehicles may be restricted in performance and not as easy to see as other vehicles. Make yourself as noticeable as possible by wearing some form of visibility aid such as a fluorescent tabard or jacket.

When will I be ready for my test?

Listen to your instructor and follow his or her advice about when to take your test.

Make sure you can demonstrate all the skills and techniques in this book and carry them out safely, without relying on your instructor for advice.

Don't come for your test before you're ready. Many people fail their test because they haven't had enough practice and experience – don't be one of them.

The theory test

All new drivers will be tested on their knowledge of *The Highway Code*. Whether you need to take a theory test will depend upon which category of test you're taking.

Category B1

Before you can book your practical driving test you'll have to pass a written theory test, unless you already hold a full licence for

- category A (motorcycles)
- category P (mopeds) obtained by passing a moped test after 1 July 1996.

The theory test is straightforward and you should pass if you're well prepared. You need to have knowledge of *The Highway Code* and understand the principles of road safety.

The test consists of 35 multiple-choice questions designed to test your knowledge of *The Highway Code* and other motoring matters. You should attempt to answer all questions, and to pass you'll have to get at least 30 right.

How do I learn the theory?

It's recommended that you study the theory at the same time as learning the practical skills.

To help you learn the theory you should study *The Highway Code*.

Other study aids include

- books
- videos
- CD-ROMs
- mock theory test papers.

The Stationery Office produce a range of study aids that are available

- from booksellers
- by calling The Stationery Office order line, see the end of this book.

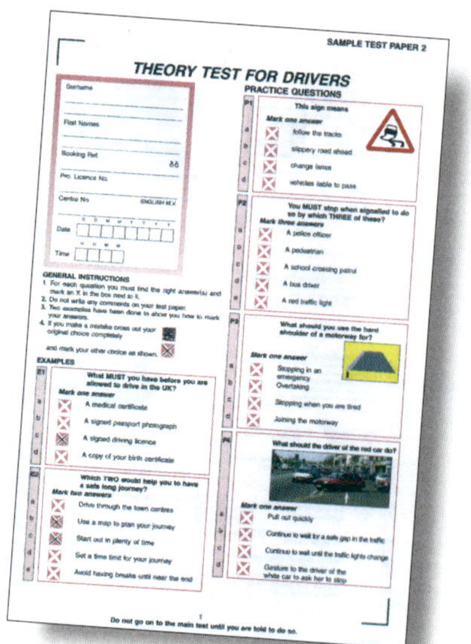

How do I book a theory test?

The easiest way to book a test is by telephone, using your credit or debit card. If you book by this method you'll be given the date and time of your test immediately. To do this call 0870 01 01 372 at any time between 8am and 6pm Monday to Friday. When you call have ready your

- driving licence number
- credit or debit card details.

If you're deaf and need a minicom machine ring 0870 01 06 372.

Welsh speakers can ring 0870 01 00 372.

You'll be given a booking number. Take this, together with your signed driving licence and photo identity, when you attend your theory test.

You may apply for your theory test by post. Application forms are available from

- theory test centres
- driving test centres
- some ADIs.

Complete the form and return it to the address shown. You must enclose the correct fee.

Don't send cash.

Categories F, G, H, K

You don't have to take a theory test for any of the vehicles in these categories, but you'll still need to

- understand and follow the rules and advice given in *The Highway Code*
- have a sound understanding of how your vehicle works, and its limitations.

Your examiner will test your knowledge by

- asking you questions on *The Highway Code* and other road safety matters
- asking you to identify some road signs
- watching how you put your knowledge into practice during your practical test.

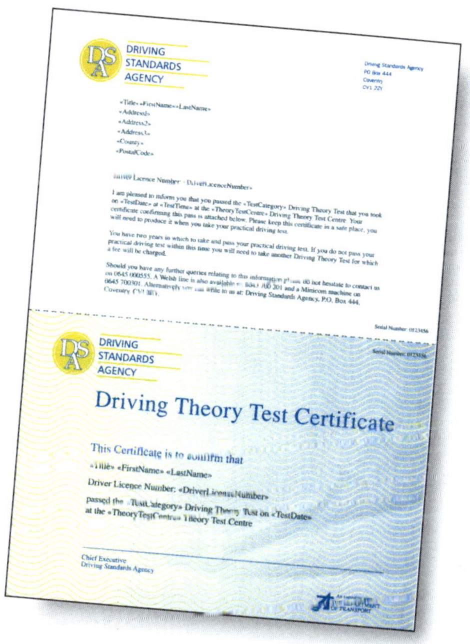

About the driving test

The driving test is straightforward. You'll pass if you can show your examiner that you can

- drive safely
- complete the set exercises
- demonstrate through your driving that you have a thorough knowledge of *The Highway Code.*

Your examiner wants you to do well and will try to put you at your ease.

How will the test be carried out?

This depends on the type of vehicle you use for the test.

Category B1 constructed like a car, with side-by-side front seats. This test will be carried out as a car test, with the examiner sitting alongside you giving instructions and watching how you drive.

Category B1 with controls and seating similar to a motorcycle. This test will be carried out as a motorcycle test. The examiner will follow you on a motorcycle or in a car, giving you instructions over a radio system that DSA provides.

When booking your test, make it clear which type of vehicle you'll be using.

Categories F, G, H, K vehicles are usually only single seat, or with a passenger seat where the examiner would not be able to watch the driver. For these tests the examiner will give you instructions at the side of the road and watch how you drive as you go around left and right circuits.

For very slow vehicles, such as pedestrian-controlled vehicles, your examiner may walk along near you where he or she can watch your driving.

Does the standard of the test vary?

No. All examiners are trained to carry out the test to the same standard.

You'll have the same result from different examiners or in different areas.

Test routes

Your examiner will use a test route which will

- include a range of typical road and traffic conditions
- take into account the size, weight and speed limitations of your vehicle.

Are the examiners supervised?

Yes, they are closely supervised. A senior officer may accompany your examiner during the test.

Don't worry about this. The senior officer won't be examining you, but making sure the examiner is testing you properly. The senior officer won't interfere with the test, so just carry on as if he or she wasn't there.

Can anyone accompany me on the test?

Yes, your instructor is allowed to be present during the test, but can't interfere or give you any help.

You should bring an interpreter with you if you need one, but you must not use an ADI for that purpose.

Your interpreter must be at least 16 years old. If you want to bring an interpreter, please make this known when you book your test.

How should I drive during the test?

Drive in the way that your instructor has taught you.

If you make a mistake, don't worry, it might be minor and may not affect your result.

Your examiner will be looking for an overall safe standard. You won't fail for one minor mistake.

How long will the test last?

About 40 minutes.

Note

The extended test for persons convicted of serious offences will last approximately twice as long.

What if I do something dangerous?

If during the test your examiner considers you to be a danger to other road users your test will stop. Make sure you're properly prepared and wait until you reach the standards set in this book before you take your test again.

What will the test include?

All practical driving tests for specialist vehicles include

- an eyesight test (if you fail this, your test won't continue)
- an emergency stop.

If you take your test on Categories B1, F or G, (fitted with a reverse gear), you'll be asked to carry out two of these three special exercises

- reversing round a corner
- turning in the road
- reverse parking.

Category H driving tests require you to drive the vehicle backwards and to turn your vehicle round, using its tracks, to face in the opposite direction. Your examiner will explain how you should perform this manoeuvre.

Category K driving tests don't include a reversing section.

Where and how are the special exercises carried out?

The special exercises will take place at carefully selected places on the test route.

Your examiner will be as helpful as possible, and for each exercise will

- ask you to pull up on the left at the side of the road
- explain the exercise and ask you to carry it out.

What if I don't understand?

Listen carefully to the explanation, but if you aren't sure about anything, ask. Your examiner understands that you may be nervous and won't mind explaining again.

What is the purpose of the test?

The driving test is designed to see if you

- can drive safely
- know *The Highway Code* and can demonstrate this through your driving.

When will I be ready for the test?

When you show that you have reached the standards set in this book – not before.

The learners who pass first time do so because they

- are well instructed
- get plenty of practice
- wait until they are ready.

Applying for your test

You can apply for a practical test by

- post
- telephone.

By post To apply by post fill out an application form, DL 26, which is available from any DSA driving test centre.

Full details of fees can be obtained from any driving test centre or DSA enquiries and booking centre; telephone 0870 01 01 372.

Complete the form and send it with the appropriate fee to the address given.

If you wish to take your test in Wales using the Welsh language, please indicate this on the form.

By phone To book your test appointment over the telephone call the DSA enquiries and booking centre; telephone 0870 01 01 372. Minicom users should telephone 0870 01 07 372 and Welsh speakers 0870 01 00 372. DSA accepts most major credit and debit cards.

Apply only when you are ready and well before the date you want to be tested.

Home tests

As the vehicles in categories F, G, H and K are often limited as to how far they can travel, it's possible for your examiner to come to you.

The booking office will send acknowledgement of receipt of your application and ask you to suggest a suitable meeting place. Your test has to take place on roads with traffic and junctions, so very remote locations may be unsuitable.

Your examiner may need to telephone you to confirm the meeting place.

Attending for your test

Make sure that you have your provisional driving licence **and** your theory test pass certificate (where necessary) with you when you arrive for your test. Check that you have signed your licence. Any of the following licences are acceptable

- a provisional driving licence issued in Great Britain (GB), or a GB full licence giving the provisional entitlement
- a provisional licence issued in Northern Ireland (NI), or a full licence giving the provisional entitlement
- an EC/EEA licence accompanied by a GB licence counterpart, if you want to take a test for a category not covered by your full EU licence.

If you have a full driving licence which was issued in another country but isn't eligible for exchange for a GB licence, you must apply for, and bring with you, a GB provisional licence. For information see DVLA information sheet D100, which is available from post offices.

Your examiner might not be able to conduct the test if you can't produce one of these licences.

If you have a photo licence you must bring with you the counterpart, which is part of the licence.

Photo identity

If your licence does not show your photograph you must also bring with you a form of photographic identification. For this your examiner will accept

- a current signed passport, which doesn't have to be a British passport
- any of the following identification cards, provided it has your photograph and your signature
  - workplace identity card
  - trade union or students' union membership card
  - card for the purchase of reduced price rail tickets
  - school bus pass
  - cheque guarantee card or credit card
  - gun licence
  - proof of age card (issued by the Portman Group)
  - standard acknowledgement letter (SAL form) issued by the Home Office

- a photograph of yourself, which has been signed and dated on the back by an acceptable person, confirming that the photograph is a true likeness of you. A signature will be accepted from the following
  - Approved Driving Instructor
  - DSA certified motorcycle instructor
  - Member of Parliament
  - local authority councillor
  - teacher (qualified)
  - Justice of the Peace
  - civil servant (established)
  - police officer
  - bank official
  - minister of religion
  - barrister or solicitor
  - medical practitioner
  - LGV trainer on DSA Voluntary Register of LGV Instructors
  - commissioned officer in HM forces.

If you don't bring these documents on the day you won't be able to take your test and you'll lose your fee. If you have any queries about what photographic evidence we will accept, contact the enquiry line, which is listed in Part Five.

No photo
No licence
No test

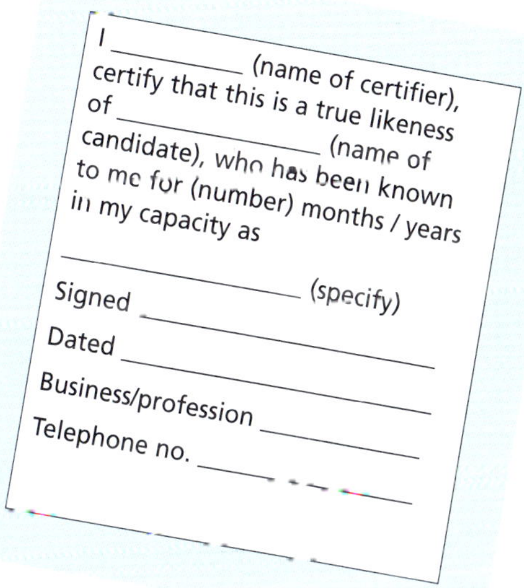
I ________ (name of certifier), certify that this is a true likeness of ____________ (name of candidate), who has been known to me for (number) months / years in my capacity as

______________ (specify)

Signed ______________

Dated ______________

Business/profession ______________

Telephone no. ______________

Your test vehicle

Make sure that the vehicle you intend to drive during the test is

- legally roadworthy and has a current MOT certificate if required
- fully covered by insurance for use on test and for you to drive.

Your examiner will ask you to sign a declaration that your insurance is in order. The test won't be conducted if you can't do so.

Your vehicle should also display

- a valid tax disc
- L plates displayed to the front and rear (in Wales, you can use D plates if you want).

L plates should **not** be displayed

- on any windscreen or back window where they could obstruct your view.
- where they will cover any lights or indicators.

If you overlook any of these

- your test may be cancelled
- you could lose your fee.

The condition of your vehicle

Your vehicle must be mechanically sound. All equipment required by law must be fitted and working correctly.

Your vehicle must also

- have clean number plates and lights
- not be carrying a load, or be partly loaded
- not be towing a trailer
- not drop mud or any debris on the road.

Unsuitable vehicles

Your examiner will also check that your vehicle is suitable for use on a practical driving test. The test cannot be conducted if the vehicle

- is operating on trade plates
- does not display a valid tax disc
- is carrying any loose items which could fall off
- has damaged lights or indicators
- has broken or missing mirrors
- is carrying any load
- could be a danger to other road users due to damage
- is fitted with an item of equipment which may cause a danger to other road users.
- has such a poor view that you need another person to help when carrying out manoeuvres or at junctions. (The driver must be able to see well enough to drive safely during the test without needing assistance from anyone else.)
- is in Category B1, where the examiner travels in the vehicle but he or she can't see clearly behind.

PART TWO THE DRIVING TEST

This part looks at the driving test itself, what your examiner will be looking for and the skills you'll need to demonstrate. It also gives advice on the problems you could face driving specific types of vehicles during your test.

The topics covered

- The eyesight test.
- Before you start the engine.
- The vehicle controls.
- Other controls.
- Moving off.
- Rear observation.
- Giving signals.
- Acting on signs and signals.
- Making progress and controlling your speed.
- The emergency stop.
- Reversing round a corner.
- Reverse parking.
- Turning in the road.
- Hazards
  - the correct routine
  - road junctions and roundabouts
  - overtaking
  - meeting and passing other vehicles
  - crossing the path of other vehicles
  - following at a safe distance
  - positioning and lane discipline
  - pedestrian crossings.
- Selecting a safe place to stop.

What the test requires

You must satisfy your examiner that, in good daylight, you can read a vehicle number plate with letters 79.4 mm (3.1 inches) high. If you need glasses or contact lenses to read the number plate you'll have to wear them throughout the test and whenever you drive.

You'll have to read the number plate at a minimum distance of 20.5 metres (about 67 feet)

You'll have to read the number plate at a minimum distance of 12.5 metres (about 41 feet)

How your examiner will test you

Before you start driving your examiner will point out a vehicle and ask you to read its number plate.

If you can't speak English or have difficulty reading, you may copy down what you see.

If your answer is incorrect, your examiner will measure the exact distance and repeat the test.

If you fail the eyesight test

If you can't show your examiner that your eyesight is up to the required standard

- you'll have failed your driving test
- your test will go no further
- you should not drive again until you have had your eyesight corrected.

Note.

If you normally wear glasses or contact lenses, always wear them whenever you drive or ride.

What the test requires

You need to make sure it's safe to start and drive your vehicle. Your examiner will watch to see that you check and adjust your seat and mirrors (if you have them). Make sure it's safe to start the engine and that you follow the correct procedures. You need to check

- that you can reach and operate the main driving controls
- that the brakes are applied
- the drive or gear selector is in neutral
- the hand throttle, accelerator or engine starting controls are properly set.

Skills and Procedures

All vehicles

- Adjust the seat.
- Adjust your mirrors.
- Check that the handbrake is applied.
- Check that the gear lever is in neutral.
- Start the engine.

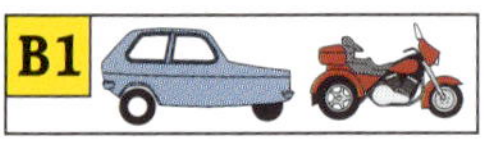

These vehicles can have motorcycle-style controls and seating or car-type controls and seating.

For motorcycle-style

Check that the fuel is turned on.

Check that the gear selector is in neutral – you may have to turn the ignition on to see any neutral light or rock the vehicle slightly to determine whether a gear is engaged .

For car-style

Check the doors are shut properly. Adjust your seat and mirrors and put on the seat belt.

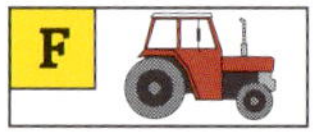

Make sure the cab doors are closed.

As well as adjusting the position of the seat, you may have to make an adjustment for your weight.

Check that the hand throttle is set correctly, if you have one.

If your vehicle has a fuel tap, make sure it's turned on.

Set the hand throttle.

If you're driving a steam roller this doesn't apply since you probably fired up an hour or two ago.

Adjust your seat. You should sit so that you're high enough to see directly, without looking through vision blocks, on military style vehicles.

Check your mirrors and adjust, if necessary. It's important to get them right, since later adjustment may not be possible.

If your vehicle has a fuel tap, make sure it's turned on.

Make sure the brakes are fully applied and latched.

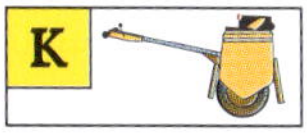

For a pedestrian-controlled vehicle, make sure your shoes are done up. A loose shoelace could be a serious hazard in heavy traffic. Check the fuel is turned on. You may have a safety lanyard which will stop the vehicle if you become separated. Make sure this is correctly in place.

Set the hand throttle or accelerator.

For a mowing machine, the blades should be disengaged and in the travelling position.

Faults you should avoid

- Starting in gear.
- Making the engine race.
- Keeping the starter motor engaged longer than necessary.
- Having to stop and readjust your seat.
- Adjusting your mirrors on the move because you didn't adjust them properly to start with.

The controls

Unlike cars, the controls of the vehicles covered by this book are not standardised, so it's important that you study and learn the layout of your vehicle thoroughly.

Advice given in this book may not apply exactly to your vehicle. Make sure you understand the differences and how they affect you.

What the test requires

You should show your examiner that you understand the functions of all the controls. You should use them

- smoothly
- correctly
- safely
- at the right time.

The main controls are

- accelerator or throttle
- clutch, if not automatic transmission
- footbrake or main brake
- handbrake
- steering
- gears.

You should

- understand what these controls do
- be able to use them competently.

Skills and Procedures

All vehicles

Accelerator and clutch

- Balance the accelerator and clutch to pull away smoothly.
- Accelerate smoothly to gain speed.
- When stopping, disengage the clutch just before you stop.
- Use the accelerator smoothly at all times.

Footbrake

- Brake smoothly and in good time.

Handbrake

- Know how and when to apply the handbrake.

Gears

- Choose the right gear for your speed and the road conditions.
- Change gear in good time.
- If you're driving an automatic vehicle, make sure the brakes are applied when you first engage 'drive' (D).

Steering

- Keep your steering movements smooth and steady.
- When turning a corner, turn accurately and at the correct time.
- Steer a steady course.

B1

Vehicles with car-style controls are covered by the 'All Vehicles' section above.

B1

Accelerator or throttle and clutch

Normally keep your hands on the grips. Only reach out towards the clutch or other controls just before you need them.

Brakes

You could have

- a front brake with a hand control and a rear brake with a foot control

or

- a linked front and rear braking system.

Use front and rear brakes together.

Parking brake

Your vehicle may have a parking brake. If not, when the engine has stopped, you may have to put the vehicle into first gear to avoid it rolling.

Steering

Keep both hands on the handlebars unless operating other controls or giving arm signals.

F

Accelerator

Use a foot-operated accelerator instead of a hand throttle or 'cruise control' for on-road driving. Hand throttles are normally for off-road work.

Footbrake

If your tractor has a split left and right braking system, use the pedals locked together. Even like this, your tractor may 'pull' to one side or the other slightly, especially if you have been working off-road using the independent brakes.

Gears

Make sure you're in the right gear range and use two-wheel drive instead of four.

Steering

Keep both hands on the steering wheel unless you're operating another control.

G

Road rollers have a variety of controls, from a conventional car-like system to a hydraulic drive system using a hand throttle to set the engine speed, and transmission controlled using a forward/reverse lever, with the brake being applied in the central neutral position.

Accelerator or hand throttle

Set the hand throttle to the correct level without the engine racing.

Brake

Brake smoothly and in good time. A heavy vehicle with steel rollers can be difficult to stop quickly.

Handbrake

Know how and when to apply the handbrake. If the brakes operate as part of the forward and reverse control, it may have a lock to keep it secured.

Gears

Choose the right gear for your speed and the road conditions. Plan your gear changes well in advance and try not to lose speed as you change.

If you're driving an automatic vehicle, make sure the brakes are applied when you first engage 'drive' (D).

Steering

Keep both hands on the wheel whenever possible even if you have a steering ball.

Control of track laying vehicles has almost standardised on a 'two stick' system of steering and braking. Occasionally you may come across a vehicle controlled by a steering wheel. If you have to drive that type of vehicle make sure you understand its controls thoroughly. Transmission is normally of the semi-automatic or fully manual gearbox type, often with high/low ranges.

Make sure you're not using the accelerator against the brakes when trying to stop in a straight line.

Clutch

When stopping, don't disengage the clutch until just before you stop. This allows the engine braking to help you to stop smoothly and in a straight line.

Accelerator

Use the accelerator smoothly at all times but especially when changing gear. You need to coordinate gear changes with the accelerator even with semi-automatic gearboxes.

Brakes

Apply the brakes evenly.

Brake smoothly and progressively. Harsh braking can make the tracks slide, especially on wet roads. As these vehicles are steered via the braking system, you should be prepared for uneven braking.

Brake latch

Know how and when to apply the brake latches, and also how to release them.

Gears

Choose the right gear and range for your speed and the road conditions.

If you're driving a semi-automatic vehicle, make sure the brakes are applied when you first engage a gear.

Steering

Keep both hands on the steering controls except when operating other controls.

On a road with a camber, tracked vehicles will wander down the slope without repeated correcting. Try to make frequent small corrections rather than occasional large ones.

When you're turning, the back of your vehicle will normally swing in the opposite direction. Allow for this when you turn.

Mowing machines and pedestrian controlled vehicles.

These may have

- automatic transmission
- two speed transmission
- a hydraulic drive controlled by a forward/reverse lever or pedal.

Throttle

If you have a hand throttle, don't let the engine race.

Clutch

The clutch should be used smoothly

If releasing the clutch applies the brakes, remember that releasing the brake will also engage the clutch.

Footbrake

Brake smoothly and in good time.

Handbrake

Know how and when to apply the handbrake. Vehicles with automatic or hydraulic drive systems can creep forward when not held securely in place.

Gears

Choose the right gear or ratio for your speed and the road conditions.

When going from forward to reverse (or reverse to forward, of course) bring your vehicle to a complete stop before engaging the opposite gear.

Steering

Keep both hands on the wheel except when operating other controls. Steering balls should be used off-road only.

The back of rear-wheel-steered vehicles will swing the opposite way when you turn. You must allow for this.

Faults you should avoid

Accelerator

- Accelerating fiercely, especially making the tyres screech or spin. This can lead to a loss of control and may distract or alarm other road users, including pedestrians.
- Making the vehicle surge or lurch by using the accelerator harshly.

Clutch

- Harsh use of the clutch when moving off or changing gear.

Footbrake

- Braking harshly, except in an emergency

Handbrake

- Applying the handbrake before the vehicle has stopped moving.
- Moving off with the handbrake on.
- On a vehicle fitted with automatic transmission, failing to set the handbrake when waiting in queues or at junctions and allowing the vehicle to creep forwards.

Gears

- Taking your eyes off the road to look at the gear lever.
- 'Coasting' with the clutch disengaged or the gear lever in neutral.

Steering

Don't turn too early when steering around a corner. If you do, you risk

- cutting the corner when turning right and putting other drivers at risk
- striking the kerb when turning left.

Don't turn too late. You could put other road users and pedestrians at risk by

- swinging wide on left turns
- overshooting on right turns.

Don't weave when travelling a steady course. If you do, you risk

- alarming oncoming drivers
- putting other road users in danger.

You should understand

The functions of all controls and switches which have a bearing on road safety, including

- lights
- indicators
- windscreen wipers
- demister
- heater.

You should know where to find these controls on the vehicle you're driving, and be able to use them without taking your eyes off the road. If, because of conditions during the test, you need to use these controls but can't, or don't because you don't know how, you may fail your test.

You need to know the meaning of any gauges or warning lights, and be able to read the speedometer accurately.

Safety checks

Although not part of the driving test, you should be able to carry out routine checks on

- oil and coolant levels
- tyre pressures
- hydraulic fluid levels.

You should also be able to identify faults with

- steering
- brakes
- tyres
- lights
- reflectors
- horn
- rear view mirrors
- speedometer
- exhaust system
- direction indicators
- windscreen wipers and washers.

What the test requires

During the test you'll have to move off from the side of the road several times. You may have to move off on a slope and from behind a parked vehicle.

How your examiner will test you

Your examiner will watch carefully to make sure that you carry this out safely and without obstructing other road users.

In particular, you must make sure that

- you check to make sure it's safe
- you signal, if it would help another road user or pedestrian
- you use the controls smoothly and safely
- you don't make other road users change speed or direction to avoid you.

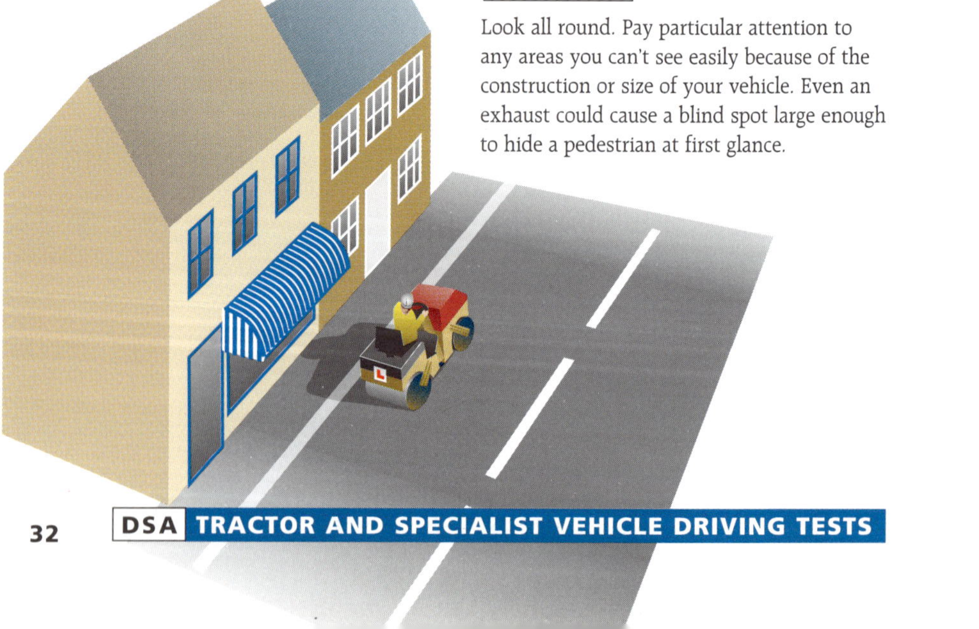

Skills and Procedures

All vehicles

- Select the correct gear.
- Check your mirrors and look all round.
- Signal, if necessary.
- If safe, move off smoothly, coordinating accelerator and clutch.

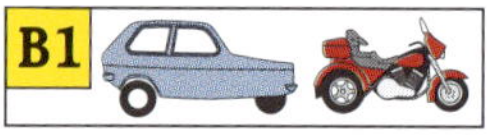

Check any blind spots before moving off – these could be caused by door pillars or even a crash helmet if you're riding a motorcycle-style vehicle.

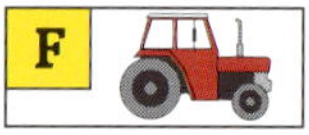

An engine cover could block your view of a child just in front of you.

Modern tractor cabs have slim frames, but these can still cause blind spots.

Look all round. Pay particular attention to any areas you can't see easily because of the construction or size of your vehicle. Even an exhaust could cause a blind spot large enough to hide a pedestrian at first glance.

H

You often have a very limited view, so make good use of your mirrors. You could even stand up and have a look round, if in doubt.

As you move off, don't turn too sharply, as the back of your vehicle may swing to the left hitting the kerb. This could damage the kerb, your tracks and endanger pedestrians. There is also the possibility that the tracks could jump off the guide wheels resulting in a serious loss of control, and damage to your vehicle.

K

You probably don't have mirrors so look all round.

If you have rear wheel steering don't turn too sharply or your wheels could hit or mount the kerb endangering pedestrians.

Faults you should avoid

- Moving off into the path of another road user.
- Stalling.
- Using the wrong gear.
- Rolling backwards on an uphill slope.
- Lurching or jerking.
- Causing another road user to change speed or direction.

Note.

Vehicles in Categories F, G, H and K may have noisy engines, so you may not be able to hear approaching vehicles.

What the test requires

During the test your examiner will be watching to see that you use your mirrors at the correct time, and act properly on what you see. There may be times when you need to look behind, as well as using your mirrors. For vehicles without mirrors you need to turn and look behind instead. Make sure you use your mirrors or look behind effectively

- before any manoeuvre
- so that you're always aware of what's happening behind you.

Check carefully before

- moving off
- signalling
- changing direction or position
- passing parked cars or obstructions
- turning to the left or right
- overtaking or changing lanes
- increasing speed
- slowing down or stopping.

If your vehicle has a driver's door opening onto the road, then you must check it's safe before opening it.

You should be aware that mirrors won't show everything behind you.

Skills and Procedures

All vehicles

- Always check your mirrors before signalling. Use the Mirror–Signal–Manoeuvre routine (see page 50). If you don't have mirrors, turn and look instead.
- Check your mirrors frequently.
- Act safely and sensibly on what you see.
- Mirrors may not show everything – turn and look when necessary.

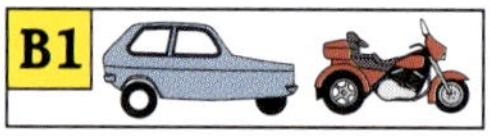

Motorcycle-style vehicles often have motorcycle mirrors, which are small and can

- be affected by vibration
- give a limited view to the rear.

Where fitted, the mirrors of these vehicles can be affected by severe vibration. If your vehicle is affected, you'll have to turn and look behind.

Track laying vehicles may have a poor rear view, which means you must rely on the mirrors. They have to be properly adjusted and clean.

Always check your mirrors before signalling. Use the Mirror–Signal–Manoeuvre routine.

Keep checking your mirrors frequently. You need to notice if a following vehicle disappears – possibly into a blind spot.

Act safely and sensibly on what you see.

These machines are not normally fitted with mirrors. Without mirrors you need to turn and look behind.

Look behind you frequently, and act safely and sensibly on what you see.

Faults you should avoid

- Signalling or changing speed or direction without checking behind first.
- Not acting correctly on what you see.
- Spending so much time looking in your mirrors that you miss things ahead.

Use of mirrors and rear observation.

Rear observation refers to a combination of mirror checks and looking behind which ensures you're always fully aware of what's happening behind you.

Looking behind is important when the view through the mirrors is restricted or blurred.

Use judgement in deciding when to look behind.

When you're looking behind you aren't looking ahead, which could be hazardous in some situations. Equally there are situations when it's dangerous not to look behind, such as before moving off.

What the test requires

During your test you'll have to give signals to let other road users know what you're going to do.

You should signal

- to let others know what you intend to do
- to help other road users, including pedestrians
- in plenty of time.

How your examiner will test you

Your examiner will watch to make sure that you signal clearly, without confusing other road users. You must only use the signals shown in *The Highway Code.*

Your signals should help other road users

- to understand what you intend to do
- to react safely.

Always make sure that your signal is cancelled after use.

Even if your vehicle has indicators and brake lights you should still know how to give arm signals, and when they may be necessary.

Skills and Procedures

All vehicles

Always check your mirrors before signalling. Use the Mirror–Signal–Manoeuvre routine.

Give signals clearly and in good time.

When using indicators, make sure you cancel them after use.

B1

For motorcycle-style vehicles, use the signals for motorcyclists shown in *The Highway Code.*

F

G

K

When travelling slowly, early signals help other road users react in good time. However, giving signals too early could mislead other road users.

If your vehicle doesn't have indicators, you'll have to use arm signals. Use only the signals shown in *The Highway Code.*

If giving arm signals, you may have to repeat them several times.

H

Make sure you cancel your indicators after use. Repeater warnings may not be easy for you to see or hear.

It's unlikely you'll be able to give arm signals when driving this type of vehicle.

Faults you should avoid

- Giving signals carelessly.
- Misleading other road users.
- Forgetting to cancel the signal.
- Waving at pedestrians to cross the road.

What the test requires

You must be able to understand

- all traffic signs
- all road markings.

You must react to them in good time.

How your examiner will test you

Your examiner will watch carefully how you react to signals and road markings. You'll be asked to turn at junctions. Make sure you act on all lane markings and direction signs.

Your examiner will not tell you about normal road signs. You must see them and act on them correctly.

Signals by other road users

- Watch for signals from other road users.
- React correctly and in good time.
- Don't rely on another's signal, it may have been left on by accident.
- Be careful how you react to signals not in *The Highway Code.* The person giving the signal might not mean what you think they mean.

Traffic lights

You must act correctly at traffic lights.

- Make sure the road is clear before proceeding at green traffic lights.
- React promptly and safely at amber traffic lights.
- Stop in the right place at red traffic lights.
- Watch for filter lights and react safely to them.

Signals given by authorised persons

You must obey the signals given by

- police officers
- traffic wardens
- school crossing patrols.

Traffic calming measures

Take extra care on roads which have been altered by the addition of

- 20 mph speed limit zones
- speed restriction humps
- width restrictions marked by bollards, posts or paved areas.

Skills and procedures

React correctly and in good time to all signals, signs and road markings.

Watch for height, width or weight restrictions that could affect you, and comply with them.

Faults you should avoid

- Blocking junctions, even if not marked with a yellow box or 'Keep Clear' markings.
- Reacting late to signals or road markings.
- Reacting incorrectly to someone else's signal.

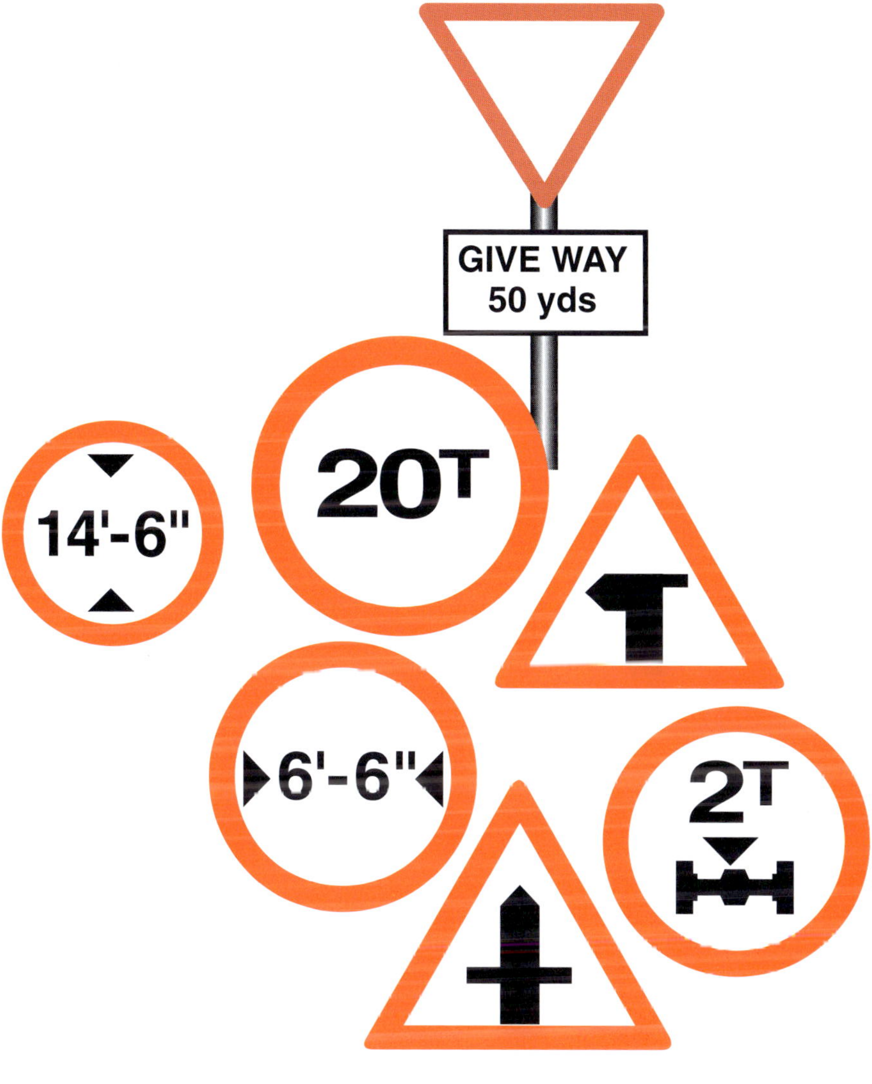
GIVE WAY
50 yds
20T
14'-6"
6'-6"
2T

What the test requires

You should make reasonable progress along the road, bearing in mind

- road conditions
- traffic
- weather
- road signs and speed limits
- the capabilities of your vehicle.

Avoid driving unnecessarily slowly.

If you're driving a slow-moving vehicle, be prepared to pull over and allow following vehicles to pass when you find a safe place to do so.

How your examiner will test you

Your examiner will be watching to make sure you control the speed of your vehicle correctly, adjusting your speed to suit the conditions at the time.

You should

- take care in the use of speed
- make sure that you can stop safely, well within the distance you can see is clear
- leave a safe distance between yourself and other vehicles
- leave extra distance on wet or slippery roads
- approach junctions and hazards at the correct speed
- make progress while keeping within
  - speed limits
  - your vehicle's limitations
  - your limitations
- show sound judgement when emerging from junctions.

Skills and Procedures

Drive at a speed appropriate to the road and traffic conditions, so that you can

- stop within the distance you can see to be clear
- react to any sudden changes in the road surface.

Faults you should avoid

- Driving too fast or too slowly for the road and traffic conditions.
- Breaking speed limits.
- Changing your speed unpredictably.
- Leaving your braking until the last possible moment.
- Using a hand throttle unnecessarily when a foot accelerator is available. If you have to use a hand throttle, make sure it is shut down when the vehicle is stationary.

What the test requires

Whenever you drive there is the possibility that something will cause you to brake as quickly as possible. You'll need to react quickly and correctly in such situations.

As part of your tractor or specialist vehicle driving test you have to show your examiner that you can stop your vehicle correctly in such an emergency.

You should be able to stop

- as quickly as possible
- safely
- under control.

How your examiner will test you

Your examiner will

- ask you to pull up at the side of the road
- ask you to stop as in an emergency when you're given the signal
- demonstrate the signal to you.

Your examiner will check that the road is clear behind you before the signal is given and will be watching carefully to see that you can stop

- in a short distance
- under full control
- without risk to other road users.

When your examiner gives the signal you should try to stop as you would in a real emergency.

You should

- react quickly
- stop in a straight line
- take account of the road conditions.

Skills and procedures

All vehicles

- React as quickly as possible.
- Brake firmly.
- Control your braking, so that you're stopping as quickly as possible without locking the wheels.
- Use both hands to control the steering.
- Just before the vehicle stops, disengage the clutch to keep the engine running.

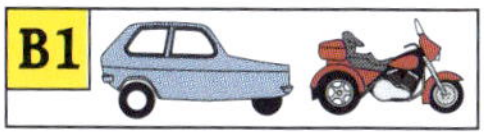

Motorcycle-style vehicles

Use front and rear brakes correctly.

Pull the clutch lever in just before you stop.

Car-style vehicles

A very narrow vehicle is in danger of turning over if it starts to skid. Ensure you control your braking properly.

The large, chunky tyres fitted to tractors can give limited grip on tarmac roads. Be prepared to take immediate correcting action if you over brake and skid, especially on wet roads.

Tractors with a split left and right braking system can veer to one side even when the brake pedals are locked together. Be prepared and take immediate correcting action if this happens.

Although not travelling very fast, the weight of a road roller can make it difficult to stop.

Metal rollers have little grip. Braking too hard for the road conditions can make even a road roller skid.

Tracked vehicles are usually stopped by applying the brakes to each track with a separate lever. You must make sure you apply the brakes evenly on each side.

If the brakes are applied unevenly, the vehicle will veer to one side.

Harsh braking can

- tip the vehicle forward, leading to loss of control
- lead to skidding, since tracks do not grip tarmac very well.

Soft tyres designed for use on grass may not grip well on wet roads.

Pedestrian controlled vehicles are usually

- slow
- light
- easy to stop. However, harsh application of the brakes could still cause skidding, especially on wet roads.

Faults you should avoid

- Anticipating the examiner's signal by slowing or stopping while your examiner is checking the road behind.
- Skidding out of control.
- Allowing the vehicle to swing off course.

What the test requires

Your examiner may ask you to reverse into a side road to demonstrate your ability to control the vehicle in reverse.

You should be able to reverse

- smoothly
- accurately
- safely
- under full control.
- with due regard for the safety of other road users.

How your examiner will test you

Your examiner will normally

- ask you to pull up just before a side road on the left
- point out the side road and ask you to reverse into it, straighten up and continue to reverse for a reasonable distance.

If your view to the rear is restricted your examiner might ask you to reverse into a road on the right.

Skills and procedures

All vehicles

- Look all round before moving off. The construction of your vehicle may cause blind areas which could hide pedestrians.
- Keep your speed down while reversing.
- Look for passing vehicles. The front of your vehicle will swing out as you turn in. Be prepared to stop if necessary.
- Continue reversing for about two vehicle lengths, keeping reasonably close to the kerb.
- Look behind you, except when checking for passing vehicles and pedestrians.

Note.
You won't be asked to carry out this manouevre if you are using a vehicle in categories

- B1 (without reverse gear)
- H
- K.

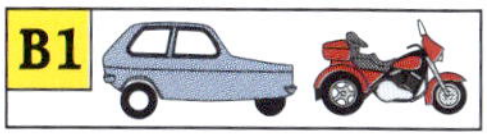

While reversing you need to be constantly aware of other road users.

You should look out of the back so you can see pedestrians or other road users approaching from behind. Avoid only looking in your mirrors.

You may be asked to reverse into a side road without a kerb. Don't let the wheels go onto the grass.

Large tractor wheels smooth out small bumps, so you may not feel the kerb if you hit or mount it.

The front of a tractor can swing a long way out when reversing into a side road. Before turning, look all round for passing traffic and stop if necessary.

Road rollers can easily damage the kerb if you let them bump or ride over it.

The noise of your engine and rollers can mask the sound of other vehicles approaching. Keep looking all round.

Faults you should avoid

- Mounting the kerb.
- Swinging out wide.
- Reversing too far from the kerb.
- Not showing consideration to other road users.
- Taking so long to complete the exercise that you create a hazard for other road users.
- Being too cautious.

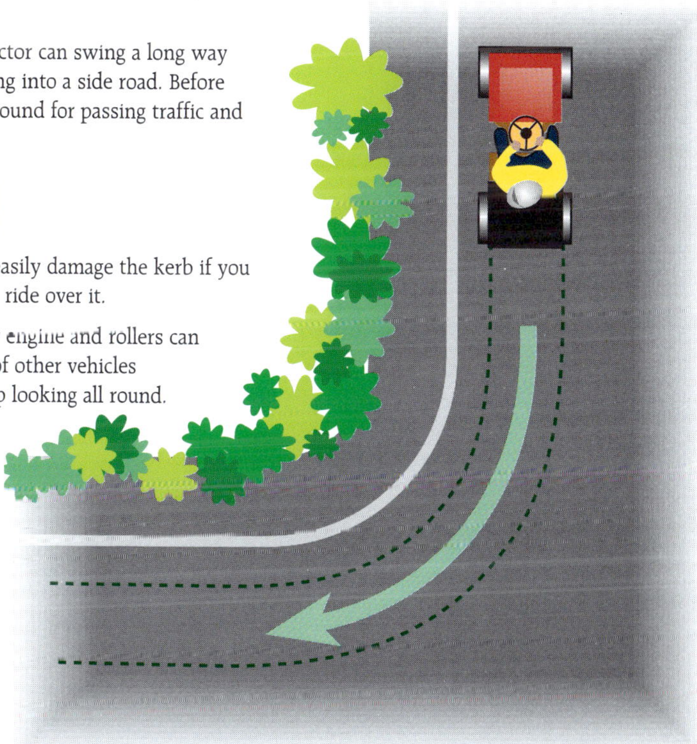

REVERSE PARKING

What the test requires

You may be asked to demostrate to your examiner that you can park your vehicle as you would when parking between two other vehicles at the side of the road. You may be asked to reverse into a space between two vehicles or behind a single vehicle, using the same method as if you were parking in a space.

You should

- be able to park your vehicle safely by reversing into a space of about twice the length of your vehicle
- stop reasonably close to, and parallel with, the kerb
- keep checking all round and avoid causing problems for other road users and pedestrians.

How your examiner will test you

Your examiner will watch carefully to make sure that you

- reverse under full control
- look all around while reversing and act on any possible danger
- complete the manoeuvre by parking close to the kerb

Note.

You won't be asked to carry out this manouevre if you are using a vehicle in categories

- B1 (without reverse gear)
- H
- K.

Skills and Procedures

All vehicles

- Before you start to reverse, have a good look all round and make sure it's safe to begin.
- Just before you turn in, look for traffic
  - ahead
  - from behind

  and stop if necessary.
- You should try to park reasonably close to the kerb.

B1

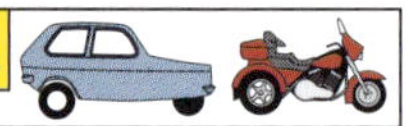

All-round observations are needed throughout this exercise.

Don't rely on your mirrors while reversing, but turn and look out of the back. This will help you see other road users as you manoeuvre.

F

Watch for the front of your tractor getting too close to the parked car. The height of your engine cover could make it difficult to judge your position.

Look for pedestrians or cyclists who may be crossing the road behind as you reverse.

G

When you're driving a large road roller, a high engine cover can make seeing the corner of the parked car difficult. Don't turn in too sharply or get too close to the kerb – if you hit or mount it your roller could easily cause damage.

Faults you should avoid

- Pulling up too close to the side of the parked car.
- Mounting the kerb.
- Parking too far from the kerb or at an angle to it.
- Reversing too far before getting close enough to the kerb.
- Causing danger to, or not showing consideration for, other road users.
- Causing an obstruction by taking an excessive length of time to park.

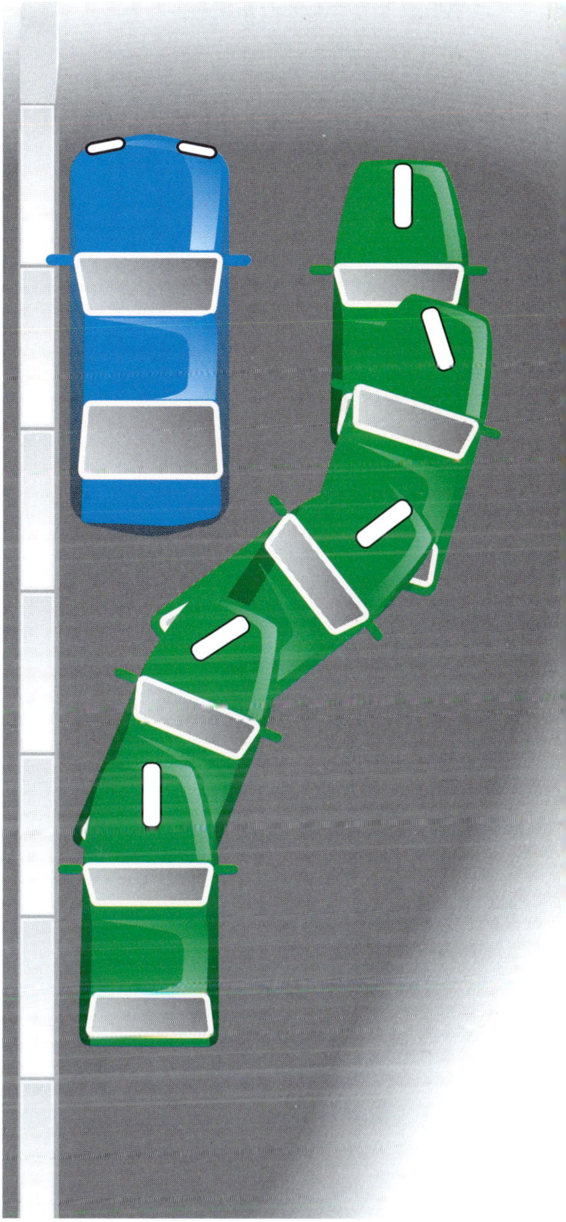

What the test requires

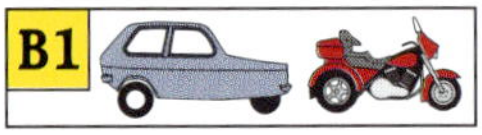

(with reverse gear fitted)

You may be asked to perform this manouvre. If you are, you must turn the vehicle so that it faces in the opposite direction, using forward and reverse gears.

- You will not necessarily be expected to complete the manoeuvre in three moves, but you should complete it in as few moves as possible.
- Keep a good look out for other road users.

You **will** be asked to perform this manoeuvre. See page 49 for details of how your examiner will test you.

How your examiner will test you

Your examiner will

- indicate a suitable place and ask you to pull up
- ask you to turn your vehicle round in the road.

Skills and Procedures

You should control your vehicle smoothly, making proper use of the

- accelerator
- clutch
- brakes
- steering.

Show awareness for other road users, including pedestrians. All-round observation is essential throughout the manoeuvre.

- Make sure that the road is clear in both directions.
- Drive forward turning to the right as much as possible.
- Steer briskly to the left just before you pull up close to the opposite kerb.
- Check all round, especially your blind spots.
- Reverse, turning to the left as much as possible.
- Steer briskly to the right before you pull up close to the kerb behind you.
- Repeat if necessary until your vehicle is facing in the opposite direction, close to the kerb.

Keep a good look out for other road users.

Look to the rear before reversing, and don't rely on mirrors while reversing.

Note.
You won't be asked to carry out this manouevre if you are using a vehicle in category

- K.

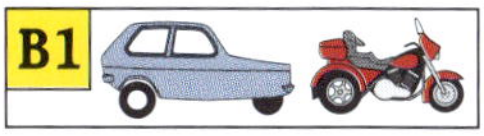

Don't rely on the mirrors. Turn and look around before beginning the manoeuvre, and keep checking all round while you're turning.

The engine covers can make it difficult to judge your distance from the kerb. Being able to judge your position accurately is an important skill, which can take time to acquire. If a road roller just clips the kerb, it can seriously damage both the kerb and the roller.

Engine noise from these vehicles can drown the sound of other vehicles. Make sure you keep checking for passing traffic.

Faults you should avoid

- Touching or mounting the kerb
- Causing danger or not showing consideration to other road users.
- Causing an obstruction for other road users by taking an excessive time to complete the exercise.

How your examiner will test category H

Your examiner will

- indicate a suitable place and ask you to pull up
- ask you to drive the vehicle backwards and cause it to face in the opposite direction by means of its tracks.

To perform this manoeuvre you may be asked to

- reverse into an opening on either the right or left and complete the turn by pulling forward
- reverse across the road and complete the turn by means of the tracks
- reverse for a distance before using the tracks to turn to face the opposite direction.

When you first move off from the side of the road, don't turn too sharply. The back of your vehicle could swing into the kerb risking damage to the kerb or your tracks.

Don't allow any part of your vehicle to overhang the pavement at any time during the manoeuvre.

When reversing you may have an extremely limited view to the rear. Go slowly, making good use of all your mirrors. If in doubt, you may even have to stop for a moment to check it's clear behind. Stand up and look if your view is poor.

How your examiner will test category B (without reverse gear)

Your examiner will

- indicate a suitable place and ask you to pull up
- ask you to ride a 'U' turn and to stop on the other side of the road.

Rear observation into the blind area is vital just before you carry out the manoeuvre.

What the test requires

A hazard is any situation which could involve adjusting your speed or altering your course.

You must take the correct action in good time to avoid danger.

How your examiner will test you

There is no specific manoeuvre for this aspect of the test.

Your examiner will watch throughout the test to see how you deal with any hazards that you come across.

Look well ahead where there are

- road junctions or roundabouts
- parked vehicles or obstructions
- cyclists or horse riders
- pedestrian crossings.

Skills and Procedures

You may have to deal with several hazards at once, or during a short space of time.

By identifying the hazard early you'll have more time to take the correct action.

Mirror – Signal – Manoeuvre (MSM routine)

Always use this routine when approaching a hazard. For vehicles without mirrors, you'll have to turn and look behind instead. Use judgement when looking round, and don't spend so much time looking behind that you miss seeing danger developing ahead.

M – Mirrors

Check the position of traffic around and behind you.

S – Signal

Signal your intention to change course or slow down. Signal in good time. Where appropriate use indicators, if your vehicle has them.

M – Manoeuvre

A manoeuvre is any change of speed or position, from slowing or stopping to turning off a busy road. This can be broken down into

- P – position
- S – speed
- L – look.

Position Your vehicle must always be in the correct position for the manoeuvre. When a change of direction is required, move into position in good time.

Due to the size of your vehicle there may be occasions when you have to straddle lane markings to avoid mounting the kerb or colliding with lamp-posts, traffic signs, etc. When positioning, watch for other road users who may not understand the special requirements of your vehicle.

Speed Ensure that the vehicle is travelling at the correct speed to complete the manoeuvre.

Look Looking means three things.

- Assessing – what you can see.
- Deciding – depending on what you see.
- Acting – either continue or wait.

Faults you should avoid

Causing other drivers to take evasive action to avoid your vehicle.

Simply looking in the mirror is not enough. You must act correctly on what you see.

What the test requires

You'll be expected to show to the examiner that you know the correct actions to take when approaching junctions and roundabouts.

You should

- use the MSM routine when you approach a junction or a roundabout
- position your vehicle correctly, adjust your speed and stop if necessary
- obey any road markings and use the correct lane. In a one-way street choose the correct lane as soon as you can do so safely.
- watch out for
  - motorbikes and cycles passing on your left
  - pedestrians crossing at or near junctions
- use effective observation before you enter or emerge from a junction.

How your examiner will test you

Your examiner will watch carefully and take account of your

- use of the MSM routine
- position and speed as you approach the hazard
- observation and judgement.

Skills and Procedures

All vehicles

- Observe road signs and markings and act correctly on what you see.
- Use the MSM routine.
- Position correctly.
- Control your speed correctly as you approach the hazard
- Slow down in good time, without harsh braking.
- Accurately judge the speed of the other traffic before joining the new road.

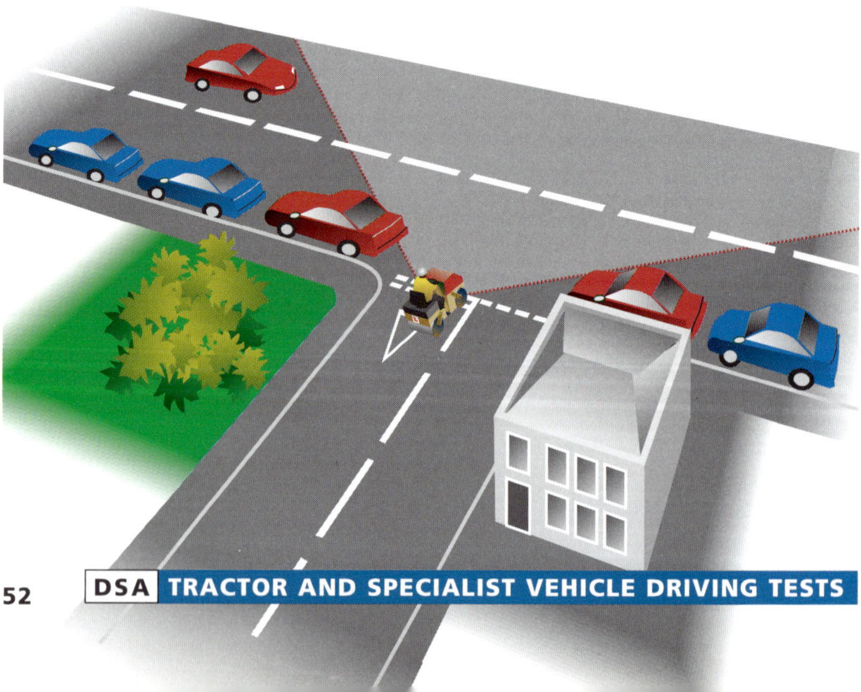

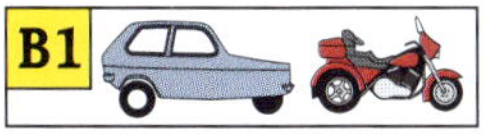

If you have a very narrow vehicle keep to the centre of the appropriate lane.

These vehicles can take up a lot of room, especially where lanes are narrow. Make sure you position yourself accurately in your lane.

You may sit a long way from the front of your vehicle. At junctions you may need to creep slowly forwards, looking both ways until you can see it's safe to emerge.

Don't leave it too late before changing lanes to put yourself in the correct position.

If you have limited speed you'll need a large gap to be able to emerge safely.

You may have a limited view to the sides. Watch out for other vehicles coming up to the junction alongside you, especially cyclists and motorcyclists.

Give clear signals. If you have to give arm signals, you may have to repeat them several times.

Make sure you pick a gap large enough to allow you to emerge safely.

Roundabouts can cause problems for slow and small vehicles. A busy roundabout with fast moving traffic can be dangerous if you can't keep up with the other vehicles. You can follow the optional advice given in *The Highway Code* for cyclists and horse riders. Keep to the left, treating each entrance and exit as a separate junction, crossing each stream of traffic when it's safe.

Faults you should avoid

- Signalling incorrectly.
- Approaching the junction at the wrong speed.
- Positioning incorrectly.
- Stopping or waiting unnecessarily.
- Entering a junction unsafely.

What the test requires

If you have to overtake any other road user, you must do so safely, allowing enough room. Give motorcyclists, cyclists and horses at least as much room as a car. They might wobble or swerve.

Allow enough space after overtaking. Don't cut in.

Don't overtake

- if the road is narrow
- where your view is limited
- where signs or road markings prohibit overtaking
- if it would stretch the performance of your vehicle to its limits.

How your examiner will test you

Your examiner will watch and take into account your

- use of the MSM routine
- reaction to road and traffic conditions
- handling of the controls.

You should be able to judge the speed and position of vehicles

- you're planning to overtake
- that might be trying to overtake you
- coming towards you.

Overtake only when you can do so

- safely
- without causing other vehicles to slow down or alter course.

Skills and Procedures

All vehicles

- Don't get too close to the vehicle ahead.
- Check that it's clear in front and behind.
- Signal.
- Give the other vehicle plenty of room.
- Check it's safe before moving back to the left.

B1 

Follow the normal guidance given in *The Highway Code.*

A small vehicle may be affected by the air flow round another vehicle. Be prepared for some buffeting.

F 

If you overtake another vehicle when driving a tractor, make sure you leave plenty of room and don't try to squeeze past.

Ensure you have enough time to overtake and return to the left without cutting in.

H 

Your vehicle may

- have very poor vision to the side and behind
- be much wider than normal.

This will make overtaking a difficult manoeuvre to perform safely. Make certain you have plenty of

- room
- time
- speed in reserve

before commiting yourself to overtaking.

G  **K** 

It would be very surprising if you had to overtake anything when driving a road roller or a pedestrian-controlled vehicle.

If you do need to move around parked vehicles, check behind before moving out and, if possible, allow enough room for a vehicle door to open. Don't move back to the left too soon.

Faults you should avoid

Overtaking when

- your view of the road ahead isn't clear
- you would have to break the speed limit.

What the test requires

You should deal with oncoming traffic safely and confidently. This applies

- on narrow roads
- where there are parked cars or other obstructions on your side of the road
- where there's an obstruction on the right and oncoming traffic has moved over to your side of the road to get past.

Use the MSM routine, and be prepared to stop and give way.

If you need to stop, keep well back from the obstruction to give yourself

- a better view of the road ahead
- room to move off easily when the road is clear.

When you're passing parked cars allow at least the width of a car door, if possible.

How your examiner will test you

Your examiner will watch carefully during the test and take into account your

- use of the MSM routine
- reaction to road and traffic conditions
- handling of the controls
- judgement and control when meeting oncoming traffic
- decision-making when moving off or stopping
- clearance when passing parked cars.

Watch out for

- doors opening
- children running out into the road
- pedestrians stepping out from between parked cars
- vehicles pulling out without warning.

Skills and Procedures

All vehicles

- Look well ahead.
- Plan where to stop to allow other vehicles room to get past.
- When you pull into a gap don't get too close to the parked cars.
- Look out for passing vehicles as you move off again.

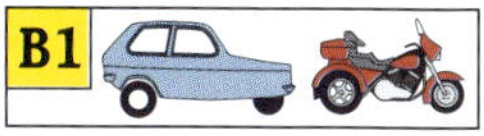

If you have a narrow vehicle, don't be tempted to drive too fast into small gaps.

Use the height of your vehicle to get a good look ahead.

Don't try and squeeze through a tight gap. Misjudging your width could have serious consequences.

If you need to wait, don't get too close to the obstruction. Engine covers or bodywork may make it difficult to judge how much clearance you have.

Ensure you have a clear view of the situation. You must accurately judge your width and length.

Watch for the back of the vehicle swinging in the opposite direction as you steer.

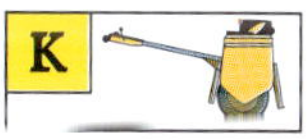

Pedestrian controlled vehicles

Remember to check for passing trafffic before moving out around an obstruction.

Mowing machines

Be aware of the width of your cutters and shrouds when moving in and out of parked cars.

For rear-wheel-steered vehicles, remember that the back of your vehicle has to move to the left for you to steer to the right, so do not get too close to parked vehicles.

Faults you should avoid

- Stopping or waiting when it's safe and correct to proceed.
- Stopping where oncoming vehicles can't get past you.
- Passing through gaps too fast for safety.

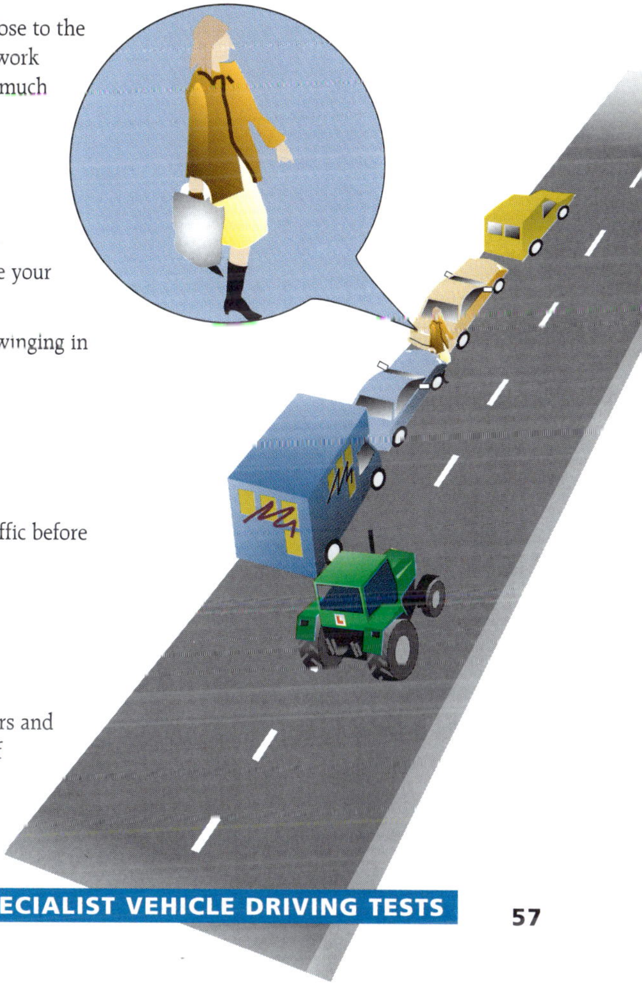

What the test requires

You should be able to cross the path of other vehicles safely and with confidence.

Crossing the path of other vehicles occurs when you have to turn right into a side road or driveway.

You should show that you can turn right safely by using the MSM routine. You must correctly judge the speed of oncoming vehicles and choose the correct time to turn.

You should

- use the MSM routine
- position your vehicle correctly and adjust your speed
- keep as close to the centre of the road as is safe
- watch out for oncoming traffic and stop if necessary.

Watch out for pedestrians

- crossing the side road which you intend turning into
- on the pavement if you're entering a driveway.

How your examiner will test you

The examiner will observe your driving and your judgement when turning right. You will be expected to show confidence and sound judgement.

Do not

- cause other road users to slow down or change direction to avoid your vehicle.

Skills and Procedures

All vehicles

- Use the MSM routine.
- Before turning right position just to the left of the centre of the road.
- Adjust your speed correctly.
- If you have to stop, stop in the correct place.
- Wait for a safe gap before turning.

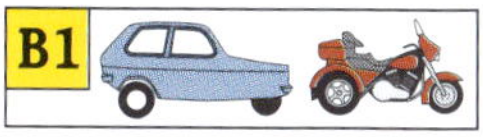

Make sure you position your vehicle correctly. Remember that your vehicle may be small and difficult for other drivers to see.

With a tractor, good vision can make positioning and judgement of oncoming traffic easier. Take your acceleration rate into account when judging a safe opportunity to turn across approaching traffic.

Wait for a large enough gap in the oncoming traffic to allow for the slow acceleration of your vehicle.

As you may be waiting for longer than a car, watch out for impatient drivers or riders trying to overtake as you turn.

As you start to turn, the back of your vehicle will swing to the left. This can be dangerous for traffic passing you on your left. Other drivers or riders may not know how your vehicle handles.

Faults you should avoid

- Causing other vehicles to
  - slow down
  - swerve
  - stop.
- Turning too early and cutting the corner.
- Going too far forward before beginning to turn.

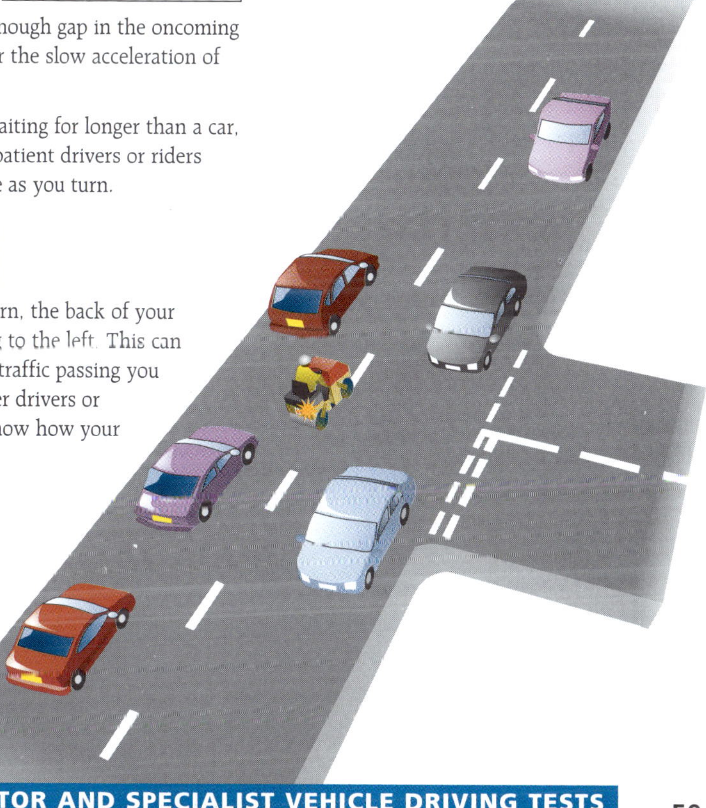

What the test requires

Keep a safe distance between your vehicle and the vehicle in front. In good conditions this should be at least 1 metre for every mile per hour you're travelling, or a two-second gap.

In poor conditions this gap should be at least doubled.

In a queue of traffic, leave enough room to pull out around the vehicle in front, in case it should break down.

How your examiner will test you

Your examiner will watch carefully and take account of your

- use of the MSM routine
- anticipation
- reaction to changing road and traffic conditions
- handling of the controls.

Skills and Procedures

All vehicles

You should

- be able to judge a safe separation distance between your vehicle and the vehicle in front
- show correct use of the MSM routine, especially before reducing speed
- avoid the need to brake harshly if the vehicle in front slows down or stops.

Take extra care when your view ahead is limited by large vehicles, such as lorries or buses.

Watch out for

- brake lights ahead
- direction indicators
- vehicles ahead braking suddenly.

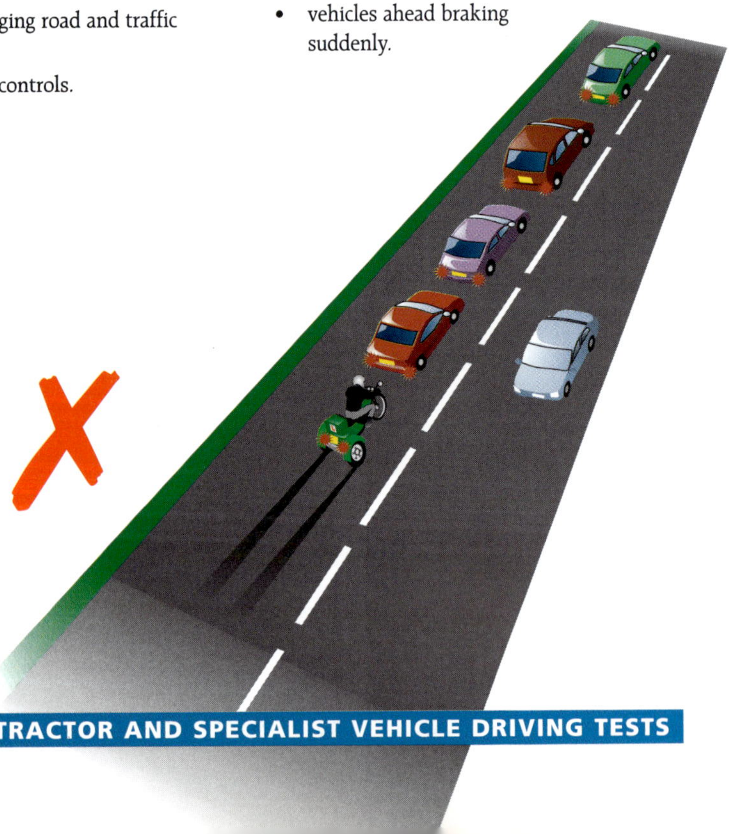

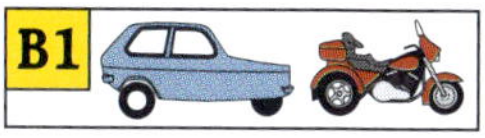

Keeping well back from the vehicle in front will give you

- a safe stopping distance in the event of an emergency
- a better view of the road ahead.

The height of a tractor can give you an excellent view of the road ahead, but longer braking distances mean you'll still have to leave plenty of room.

Unless you're in slow-moving traffic queues, you're not going to be close to traffic ahead.

In a traffic queue, look well ahead and don't just react to the vehicle in front. Leave plenty of room, and try to brake in good time. Steel rollers can skid on wet roads if you brake hard and late.

The height of your vehicle helps by allowing you a good view of the road ahead.

In traffic queues keep a safe distance from the vehicle in front. This will allow you to pass the vehicle in front if it stops or breaks down.

Faults you should avoid

- Following too closely.
- Braking suddenly.
- Stopping too close to the vehicle in front in a traffic queue.

What the test requires

Correctly positioning your vehicle is an important part of driving. A badly-positioned vehicle causes difficulties and confusion for other drivers. Normally, keep well to the left.

You should

- keep clear of parked vehicles
- avoid weaving in and out between parked vehicles
- position your vehicle correctly for the direction you intend to take.

You should obey all lane markings, especially

- left or right-turn arrows at junctions
- when approaching roundabouts
- in one-way streets
- bus and cycle lanes.

How your examiner will test you

Throughout the test the examiner will observe your road position. You will be expected to select the correct position in good time for all situations.

Skills and Procedures

All vehicles

- Plan ahead and choose the correct lane in good time.
- Use the MSM routine correctly.
- Follow any road markings as soon as you can.
- Position your vehicle correctly, even if there are no road markings.

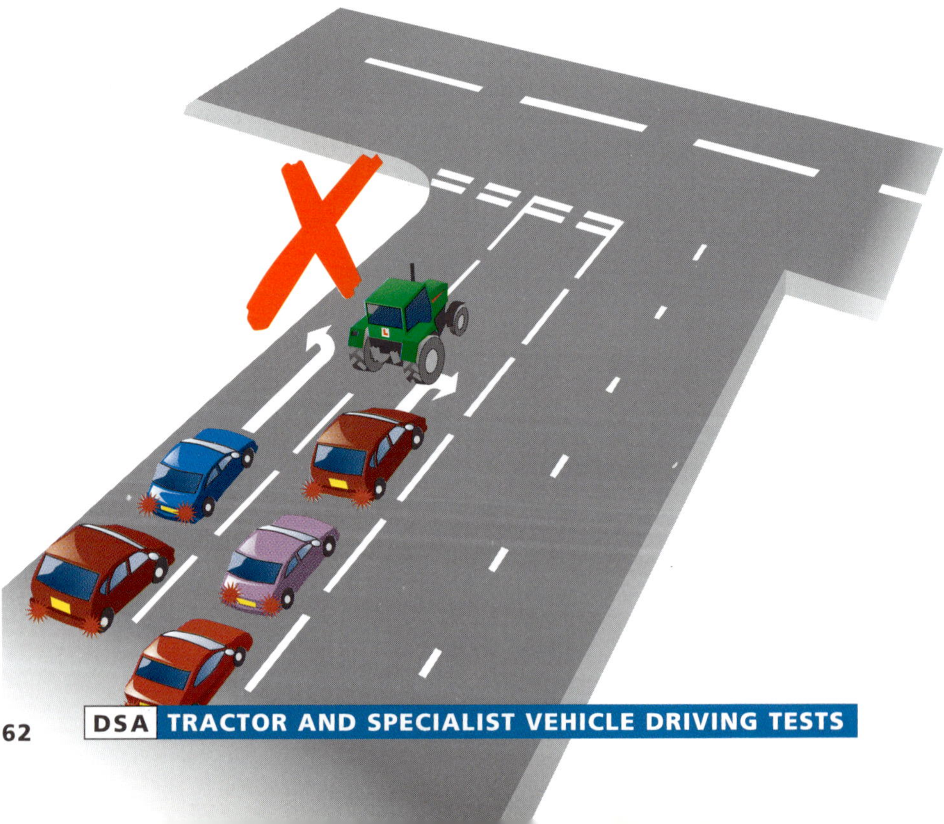

B1 

Position your vehicle in the middle of your lane. Incorrect positioning could encourage another driver to pass unsafely.

F  **G** 

The height of these vehicles can make it easier to see and plan ahead. Road markings, especially, can be easier to see.

Get into your lane as soon as you can. The slower you're travelling, the harder it becomes to change lanes safely.

No matter how slowly you travel, bus and cycle lane restrictions will apply to you.

H 

A wide vehicle can take up a whole lane. Try to keep to the middle of the lane allowing other vehicles as much room as possible to get past.

K 

Make yourself as visible as possible by wearing high visibility clothing.

At busy roundabouts pedestrian-controlled vehicles may find it safer to adopt the procedure for cyclists in *The Highway Code*, Rule 163.

Remember that bus and cycle lane restrictions will apply to you.

Faults you should avoid

- Driving too close to the kerb.
- Driving too close to the centre of the road.
- Changing lanes at the last moment.
- Obstructing other road users by being badly positioned or being in the wrong lane.
- Straddling lanes or lane markings.
- Cutting across the path of other traffic at roundabouts.

What the test requires

Pedestrian crossings are only safe if drivers treat them correctly.

The different types of pedestrian crossings are

- zebra crossings
- pelican crossings
- puffin crossings
- toucan crossings.

At pelican, puffin and toucan crossings you must

- stop if the lights are red.

At a pelican crossing, you should

- give way to any pedestrians when the amber lights are flashing.

At a toucan crossing, you should

- give way to cyclists as you would to pedestrians.

How the examiner will test you

Your examiner will watch carefully and take account of how you deal with all types of pedestrian crossings.

You should be able to

- control your speed as you approach a pedestrian crossing
- stop safely if necessary
- move off when it's safe.

Skills and Procedures

All vehicles

- On the approach to a pedestrian crossing use the MSM routine.
- Approach so that you can stop easily if necessary.
- Before you move off, look for pedestrians hurrying onto the crossing.

B1 

You should know when and how to give the arm signal for slowing and stopping. This can help

- pedestrians
- drivers of following vehicles
- drivers of approaching vehicles.

F  G 

H 

Large, noisy vehicles can be intimidating to pedestrians, especially children and the elderly. Approach with care. If you stop, wait patiently – the pedestrians may not react instantly.

K 

Pedestrians may misjudge the speed of these vehicles. Be prepared for someone to step out at the last moment.

Faults you should avoid

- Approaching a crossing too fast.
- Driving over a crossing without stopping or showing awareness of waiting pedestrians.
- Blocking a crossing by stopping directly or partly across it.

Don't hurry pedestrians by

- sounding the horn
- revving the engine
- edging forward.

Don't

- overtake within the zigzag white lines leading up to crossings
- wave pedestrians off the pavement
- react late or incorrectly to traffic light signals at controlled crossings.

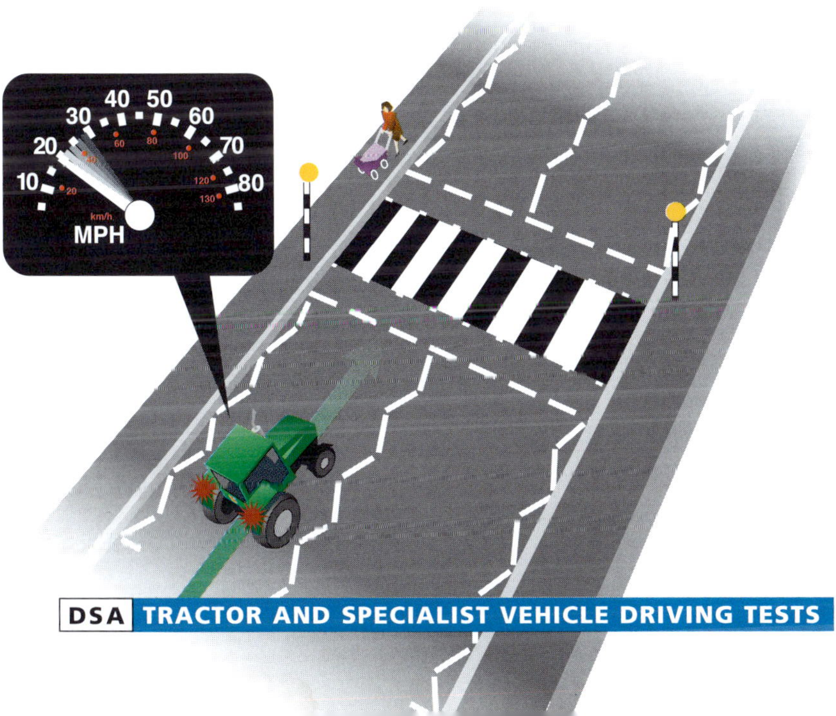

What the test requires

Whatever vehicle you drive, you'll have to stop at the side of the road safely and without inconvenience to other road users. Normally, your examiner will leave it up to you where you stop. Occasionally you'll be asked to stop in a specific place.

When stopping you should be able to select a place where you won't

- obstruct the road
- create a hazard
- break the law.

How your examiner will test you

Your examiner will watch and take account of your

- use of the MSM routine
- judgement in selecting a safe place to stop.

You should know how and where to stop without causing inconvenience or danger to other road users.

Skills and Procedures

All vehicles

- Use the MSM routine.
- Select a safe place to stop.
- Brake and steer smoothly.
- Stop close to, and parallel with, the kerb.
- Normally, stop with your wheels straight.

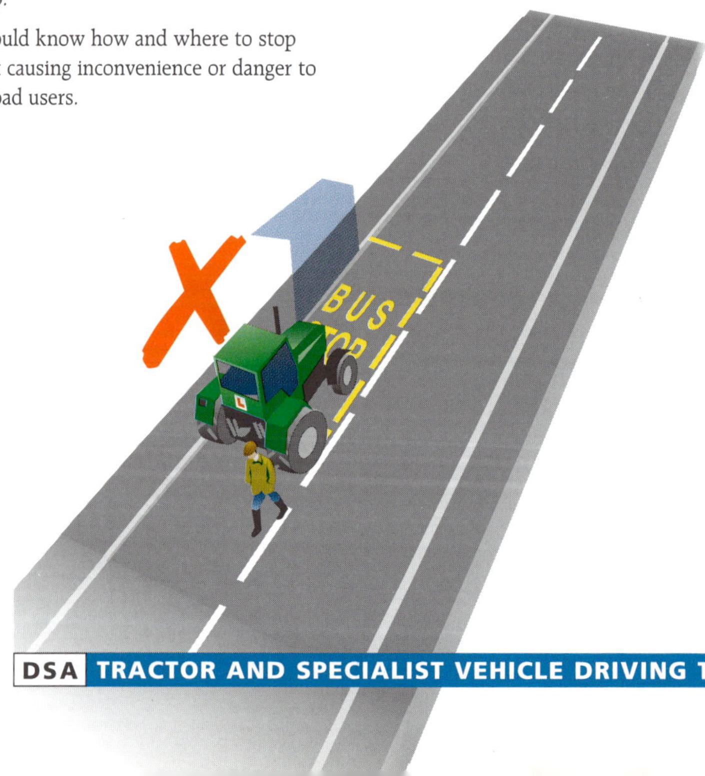

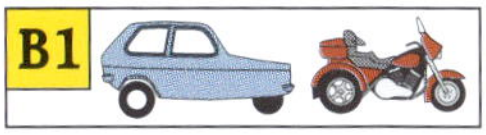

When parking in marked-out spaces, park neatly and don't straddle the markings.

If you have a high cab, watch out for low tree branches or street furniture.

If you hit or mount the kerb you may cause damage.

The rear of a tracked vehicle will swing towards the kerb as it moves off. Make allowance for this when you stop.

Mowing machines sometimes have cutter guards which increase the overall width. Don't allow these guards to overhang the pavement when you stop.

Faults you should avoid

- Stopping without sufficient warning to other road users.
- Causing danger or inconvenience to other road users when you stop.
- Stopping anywhere *The Highway Code* says you shouldn't.

The topics covered

- If you pass.
- Developing your driving standards.
- If you don't pass.
- DSA complaints guide for test candidates.
- DSA compensation code for test candidates.

If you pass

Congratulations! Passing your driving test shows that you can drive safely.

You'll be given

- a pass certificate (D10)
- a copy of the driving test report, which will show driving faults which have been marked during the test
- a debrief. The examiner will discuss any driving faults marked which need to be drawn to your attention
- guidance notes.

Look carefully at the report with your instructor. This is given to you to help you overcome any weaknesses in your driving.

You should apply for your full licence as soon as possible. Follow the instructions on the back of your pass certificate (D10).

If you don't apply within two years, you'll lose your full licence entitlement. You'll then revert to learner status.

Developing your driving standards

Passing your driving test is the first step in becoming a good driver. It means you have reached a safe basic standard.

You should aim to raise your standard of driving with additional instruction and experience.

Motorway driving

If you can now drive a vehicle on motorways, you must make sure you understand the rules and regulations for these roads. Motorway driving calls for new skills and judgement. Your instructor may be able to assist you in gaining some experience before you drive on your own.

The Highway Code

Own a current copy of *The Highway Code* and refer to it often. It will keep you up to date with developments in driving rules and procedures.

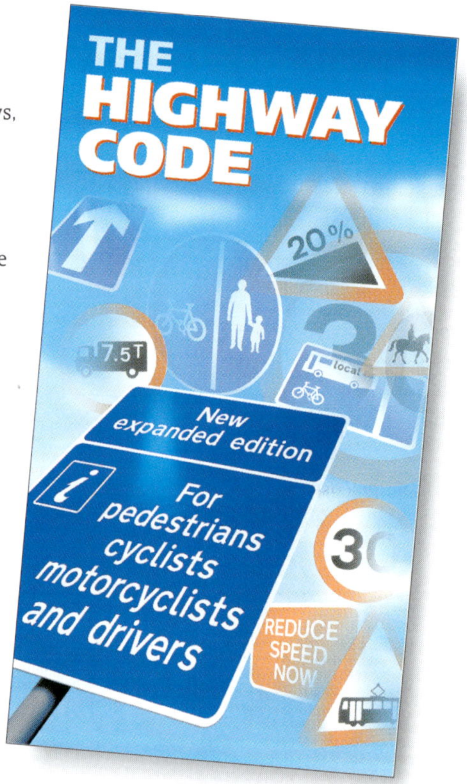

If you don't pass

Your driving isn't yet up to the required standard. You'll have made mistakes which have shown a lack of control or could cause danger on the road.

Your examiner will help by giving you

- a driving test report form. This shows all the faults marked during the test
- a brief explanation of why you haven't passed
- guidance notes.

Your examiner will be able to help you by pointing out the aspects of your driving which you need to improve.

Study the driving test report and refer to the relevant sections in this book.

Show your copy of the report to your instructor, who should advise and help you to correct the faults. Listen to your instructor's advice carefully, and get as much practice as you can.

Don't just practise the things you have failed for, try to improve all aspects of your driving.

Right of appeal

You'll obviously be disappointed if you don't pass your driving test. Although your examiner's decision can't be changed, if you think your test wasn't carried out according to the regulations, you have the right to appeal.

If you live in England and Wales you have six months after the issue of the Statement of Failure in which to appeal (Magistrates Courts Act 1952 Ch. 55 part VII, Sect. 104).

If you live in Scotland you have 21 days in which to appeal (Sheriff Court, Scotland Act of Sederunt (Statutory Appeals) 1981).

Before lodging an appeal you should seek legal advice.

DSA COMPLAINTS GUIDE FOR TEST CANDIDATES

DSA aims to give its customers the best possible service. Please tell us

- when we have done well
- when you're not satisfied.

Your comments can help us to improve the service we have to offer. For information about the DSA service standards, contact your local Area Office.

If you have any questions about how your test was conducted, please contact the local Supervising Examiner, whose address is displayed at your local test centre.

If you are dissatisfied with the reply, or if you wish to comment on other matters, you can write to DSA.

If your concern relates to an ADI you should write to

The Registrar of
Approved Driving Instructors
Driving Standards Agency
Stanley House
56 Talbot Street
Nottingham NG1 5GU.

Finally, you can write to

The Chief Executive
Driving Standards Agency
Stanley House
56 Talbot Street
Nottingham NG1 5GU.

None of this removes your right to take your complaint to

- your Member of Parliament, who may decide to raise your case personally with the DSA Chief Executive, the Minister, or the Parliamentary Commissioner for Administration (the Ombudsman), whose name and address listed in Part Five.
- A magistrates court (in Scotland, the Sheriff of your area) if you believe that your test wasn't carried out according to the regulations.

Before doing this you should seek legal advice.

DSA always aims to keep test appointments, but occasionally we have to cancel a test at short notice. We will refund the test fee, or give you your next test free, in the following circumstances

- if we cancel a test
- if you cancel a test and give us at least ten working days' notice
- if you keep the test appointment, but the test doesn't take place or isn't finished, for a reason that isn't your fault or the fault of the vehicle you're using.

We will also compensate you for any money you lose because we cancel your test at short notice (unless it was for bad weather). For example, we will pay

- the cost of hiring a vehicle for the test, including reasonable travelling time to and from the test centre
- any pay or earnings you lost, after tax and so on (usually for half a day).

We WON'T pay the cost of driving lessons which you arrange linked to a particular test appointment, or extra lessons you decide to take while waiting for a rescheduled test.

How to apply

Please write to the DSA Enquiries and Booking Centre and send a receipt showing vehicle hire charges, or an employer's letter which shows what earnings you lost. If possible, please use the standard form (available from every driving test centre or booking office) to make your claim.

These arrangements don't affect your legal rights.

PART FOUR DRIVING LARGE HEAVY OR SLOW VEHICLES

This part looks at the responsibilities you accept once you've passed your test and become a driver.

The topics covered

- Planning your journey.
- Amber beacons.
- Loads.

Planning your journey

When you have passed your test you may be driving large or slow vehicles on a variety of roads. Try and plan your route to avoid places where you could cause congestion and inconvenience to other road users.

Try to avoid

- busy roads
- known road works
- travelling during busy times of the day
- narrow roads with no passing places.

If you're causing a queue of traffic behind, follow the advice in *The Highway Code*. Pull over where it's safe to do so and let following traffic pass.

Height, width and weight limits

You need to know the size and weight of your vehicle. Before you start your journey, plan your route and ask for advice if you don't know the area. Local Authorities will know of any

- height restrictions
- width restrictions
- weight restrictions.

Motoring organisations may also be able to help if travelling further afield.

At level crossings follow the advice given in *The Highway Code*.

Motorways

Motorways must not be used by

- holders of provisional ordinary licences
- pedestrians
- animals
- riders of motorcycles under 50 cc
- cyclists and horse riders
- agricultural vehicles
- certain slow-moving vehicles carrying oversized loads (except by special permission).

Amber beacons

On an unrestricted dual carriageway any vehicle with four or more wheels must display an amber beacon if it cannot exceed 25 mph.

The beacon must be in a position where it can clearly be seen by other road users. If a trailer is being towed which causes the beacon on the towing vehicle to be obscured, a beacon must be displayed on the rear of the trailer.

A beacon is not required if the vehicle is only crossing the dual carraigeway.

Dirty vehicles

Many of the vehicles covered by this book will be used in fields or on construction sites and road works. Before leaving the site you should check the vehicle to make sure it's safe and legal to take onto the road.

You should check that

- indicators and lights are clean
- number plates are clean and can be read easily
- windscreens and windows are clean
- the vehicle will not drop mud or any material on the road. Many sites have inspection and cleaning facilities near the exit. Make sure you use them.

Loads

Any load which is carried on your vehicle or trailer must be secure. If you intend to carry a large load you should seek advice from the local police.

When loading any vehicle or trailer, the following points should be considered.

- Make sure the vehicle or trailer is capable of carrying the load.
- Keep any load beds clean and dry. This will help stop the load sliding.
- When carrying a small load, put it against the headboard.
- Always secure the load.
- If the bodywork of the vehicle or trailer is being used to contain the load, make sure it is strong enough to do so.
- Check there is no leakage.
- If you are using an open back vehicle or towing a trailer, use a sheet or a tarpaulin.
- Don't allow anything to blow off in the wind. Even dust from a vehicle can cause injuries to pedestrians.
- If you have to stack items, make sure the largest and heaviest are on the bottom.
- Make sure all anchorage points are secure and free of rust.
- Ensure any ropes you use are dry and in a good condition.
- If the load has a large overhang either to the side or the rear, then you must put warning signs or lights on it.
- If the load covers the vehicle or trailer lights, then additional lights must be displayed.

Sharp edges

It's the driver's responsibility to make sure that the vehicle does not endanger any other road user. There must not be any sharp edges or projections.

Agricultural impliments, such as a ploughs, have sharp edges. These must be protected and marked so that they are clearly visible to pedestrians.

Before being used on the road, mowing machines must have the blades stowed in the travelling position.

PART FIVE ADDITIONAL INFORMATION

This part lists

- DSA addresses.
- Other addresses you may find useful.

DSA addresses

DSA Test Enquiries and Booking Centre

DSA
PO Box 280
Newcastle-upon-Tyne
NE99 1FP

Tel:	0870 01 01 372
Welsh speakers:	0870 01 00 372
Minicom:	0870 01 07 372
Fax:	0870 01 02 372

DSA Head Office

Stanley House
56 Talbot Street
Nottingham
NG1 5GU

Tel: 0115 901 2500
Fax: 0115 901 2940

Other addresses you may find useful

Approved Driving Instructors' National Joint Council

The Secretary
41 Edinburgh Road
Cambridge CB4 1QR

Tel: 01223 359 079

Driving Instructors Association

The Secretary
Safety House
Beddington Farm Road
Croydon CR0 4XZ

Tel: 0181 665 5151

Motor Schools Association of Great Britain Ltd

The General Manager
182A Heaton Moor Road
Stockport
Cheshire SK4 4DU

Tel: 0161 443 1611

The Parliamentary Commissioner for Administration (The Ombudsman)

Mr M Buckley
Millbank Tower
Millbank
London SW1P 4QP

Tel: 0171 217 4163

Driver and Vehicle Licensing Agency DVLA

Customer Enquiry Unit
Swansea SA6 7J

Tel: 01792 772151

Lantra

National Agricultural Centre
Kenilworth
Warwickshire
CV8 2LG

Tel: 01203 696996

National Farmers Union

Agriculture House
164 Shaftesbury Avenue
London
WC2H 8HL

Tel: 0171 331 7200

Fax: 0171 331 7313

National Farmers Union (Scotland)

Rural Centre
West Mains
Ingliston
New Bridge
Midlothian EH28 8LT

Tel: 0131 472 4000

Fax: 0131 472 4010

Essential guidance for all types of driving available from the Driving Standards

The Highway Code

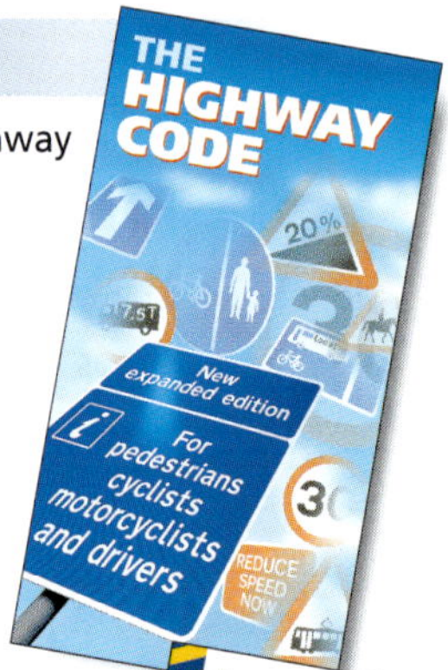

Research has proved conclusively that keeping a copy of The Highway Code in the home significantly improves driver knowledge, and therefore safety, on the road. The latest Highway Code has been completely revised to include new legislation, with greater emphasis on safe driving and special sections on vulnerable road users such as horse riders, cyclists and the elderly. Plus advice to reflect the changes in lifestyle and technology, with rules dealing with driver fatigue and recommendations about the use of mobile phones.

ISBN 0 11 551977 7 £1.49

The Official Theory Test for Car Drivers and Motorcyclists

This best selling title is the official publication – written and compiled by the DSA to help you to pass the theory test. Covering all aspects of the test, it is the most comprehensive guide available, and includes the very latest information, with the new enlarged question bank of around 1,000 questions in an easy to use format.

ISBN 0 11 552196 8 £11.99

Test Yourself Papers for the Driving Theory Test

Five practice papers for the theory test, with questions taken from the official question bank and the correct answers supplied for further reference and revision. Be fully prepared by simulating the theory test at your leisure.

ISBN 0 11 551984 X £4.99 (INCL VAT)

The Official Driving Test

The practical driving test – fully illustrated and written in a clear and easy-to-understand style.

Fully updated with the officially recommended syllabus for learning to drive.

ISBN 0 11 552190 9 £6.99

and driving instruction: Agency and The Stationery Office

Know Your Traffic Signs

This useful publication illustrates and explains the vast majority of traffic signals, signs and road markings which any road user is likely to encounter. It is the most comprehensive explanation of the signing system available, and is exceptional value for money.

ISBN 0 11 551612 3 £2.50

Official DSA Guide for Driving Instructors

Defining the syllabus that all new, existing or aspiring ADIs need to follow. This title explains all the aspects of driving and driving instruction which are assessed during each of the stages, and also gives an insight into the marking system used. Written by the DSA – who administer and set the standards for the register of Approved Driving Instructors (Car) – the ADI Register – it explains what is required for entry on to the register, and the subsequent checks of continuing ability.

ISBN 0 11 551785 5 £18.99

The Official Compulsory Basic Training for Motorcyclists

Updated to include the latest definitive advice and all the information for the first step for motorcycle learners. CBT is the course that you must complete before riding with a provisional licence. This title, written and compiled by the DSA explains each of the various elements of the training, with helpful information for pre and post course assistance.

ISBN 0 11 552192 5 £6.99

The Official Motorcycling Manual

The authoritative guide to motorcycling, containing new graphics and photographs and updated to include the very latest information and advice. An essential reference book for motorcyclists who are keen to improve their machine handling and safety skills, including expert guidance on riding in traffic, defensive driving, in bad weather and at night.

ISBN 0 11 552193 3 £9.99

Official Theory Test for Drivers of Large Vehicles

Complete theory test instruction for drivers of LGVs, buses and coaches. With all the questions you are likely to be asked, reasoned explanations and advice on where and when the test can be taken.

ISBN 0 11 552168 2 £12.99

Bus and Coach Driving Manual

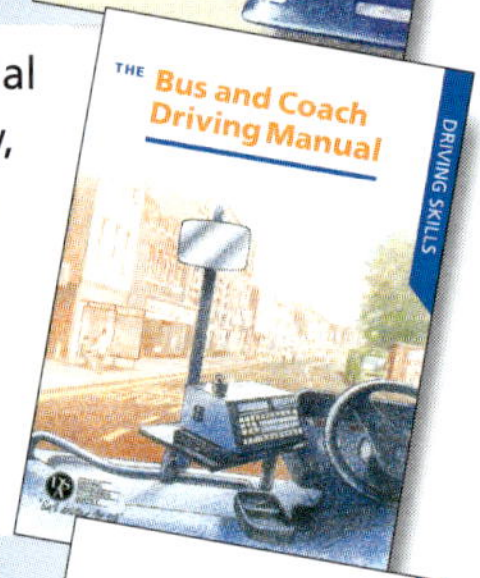

Definitive information on the PCV practical test: everything from initial applications and the latest legislation, through to general road safety, driving attitude, First Aid and, of course, the officially recommended syllabus. A companion volume to the *Official Theory Test for Drivers of Large Vehicles*.

ISBN 0 11 551784 7 £12.99

Goods Vehicle Driving Manual

All the necessary information to help you to pass the LGV practical test. An alternative companion volume to the *Official Theory Test for Drivers of Large Vehicles*, it combines the official recommended syllabus with sound advice.

ISBN 0 11 551783 9 £12.99

Official Goods Vehicle Driving Video

Safe driving practice and defensive driving techniques for both new and established learner drivers. Covering all classes of goods vehicle above 3.5 tonnes, this video stresses how good, all round observation and sound planning can reduce accident risk. With an approx running time of 54 minutes, it offers a realistic approach to the realities of meeting tight schedules whilst taking maximum care of other road users, drivers and vehicles.

ISBN 0 11 551971 8 £29.99 (INC VAT)

The Stationery Office publishes a wide range of titles on behalf of the Driving Standards Agency covering all types of driver training – from CBT for motorcyclists and the theory and practical tests for new road users, through LGV and PCV guidance and on to tractors and specialist vehicles. For further information please contact our general enquiries: 0171 873 0011 or visit us Online: www.tso-online.co.uk

Printed in The United Kingdom for The Stationery Office J0079413 06/99 C500 63789

Audio Mock Theory Test

Devised to help new drivers prepare for the Theory Test with simulated test conditions. Just ring

0930 217 217

The test takes 30 minutes to complete, callers must be 18 or over, and calls are charged at 50p per minute.

Cyberdrive

The website for new drivers – aimed to help you to pass your test and advance your driving skills:

www.cyberdrive.co.uk

Order Form

These titles are available from:

- **The Stationery Office, The Publications Centre, PO Box 276, London SW8 5DT**
  Mail or Fax this order form, or telephone.
  Telephone orders: 0171 873 9090 quoting ref ACJ/TSV
  Fax orders: 0171 873 8200
  General enquiries: 0171 873 0011
  Online: www.tso-online.co.uk

  Stationery Office publications are also available from:
- **The Stationery Office Bookshops**
  (see inside back cover for details)
- **Accredited Agents and all good booksellers and multimedia retailers**

Please send me the following titles:

Title	*ISBN*	*Quantity*	*Price*
The Highway Code	ISBN 0 11 551977 7	....	£1.49
The Official Theory Test for Car Drivers and Motorcyclists	ISBN 0 11 552196 8	....	£11.99
Test Yourself Papers for the Driving Theory Test	ISBN 0 11 551984 X	....	£4.99 (incl VAT)
The Official Driving Test	ISBN 0 11 552190 9	....	£6.99
Know Your Traffic Signs	ISBN 0 11 551612 3	....	£2.50
The Official DSA Guide for Driving Instructors	ISBN 0 11 551785 5	....	£18.99
The Official Compulsory Basic Training for Motorcyclists	ISBN 0 11 552192 5	....	£6.99
The Official Motorcycling Manual	ISBN 0 11 552193 3	....	£9.99
Official Theory Test for Drivers of Large Vehicles	ISBN 0 11 552168 2	....	£12.99
Bus and Coach Driving Manual	ISBN 0 11 551784 7	....	£12.99
Goods Vehicle Driving Manual	ISBN 0 11 551783 9	....	£12.99
Official Goods Vehicle Driving Video	ISBN 0 11 551971 8	....	£29.99 (incl VAT)
	Handling charge per order		£2.94
	Total enclosed		£.........

PLEASE COMPLETE IN BLOCK CAPITALS

Name ....

Address ....

....

.... Postcode.................. ACJ/TSV

☐ I enclose a cheque for £.................. payable to: *'The Stationery Office'*

☐ Please charge to my account with The Stationery Office, No:

....

☐ Please debit my Mastercard/Visa/Amex/Diners/Connect Card Account No.

Signature.. Expiry date ....

☐ Please tick if you do not wish to receive further details on titles from The Stationery Office.
Prices are correct at time of going to press but may be subject to change without notice. Stationery Office Account holders should note that credit card transactions will not appear on their statements. A full listing of terms and conditions of sale can be obtained on request from The Stationery Office Publications Centre, 51 Nine Elms Lane, London SW8 5DR.

The Stationery Office

DSA DRIVING STANDARDS AGENCY SAFE DRIVING FOR LIFE

NOTES